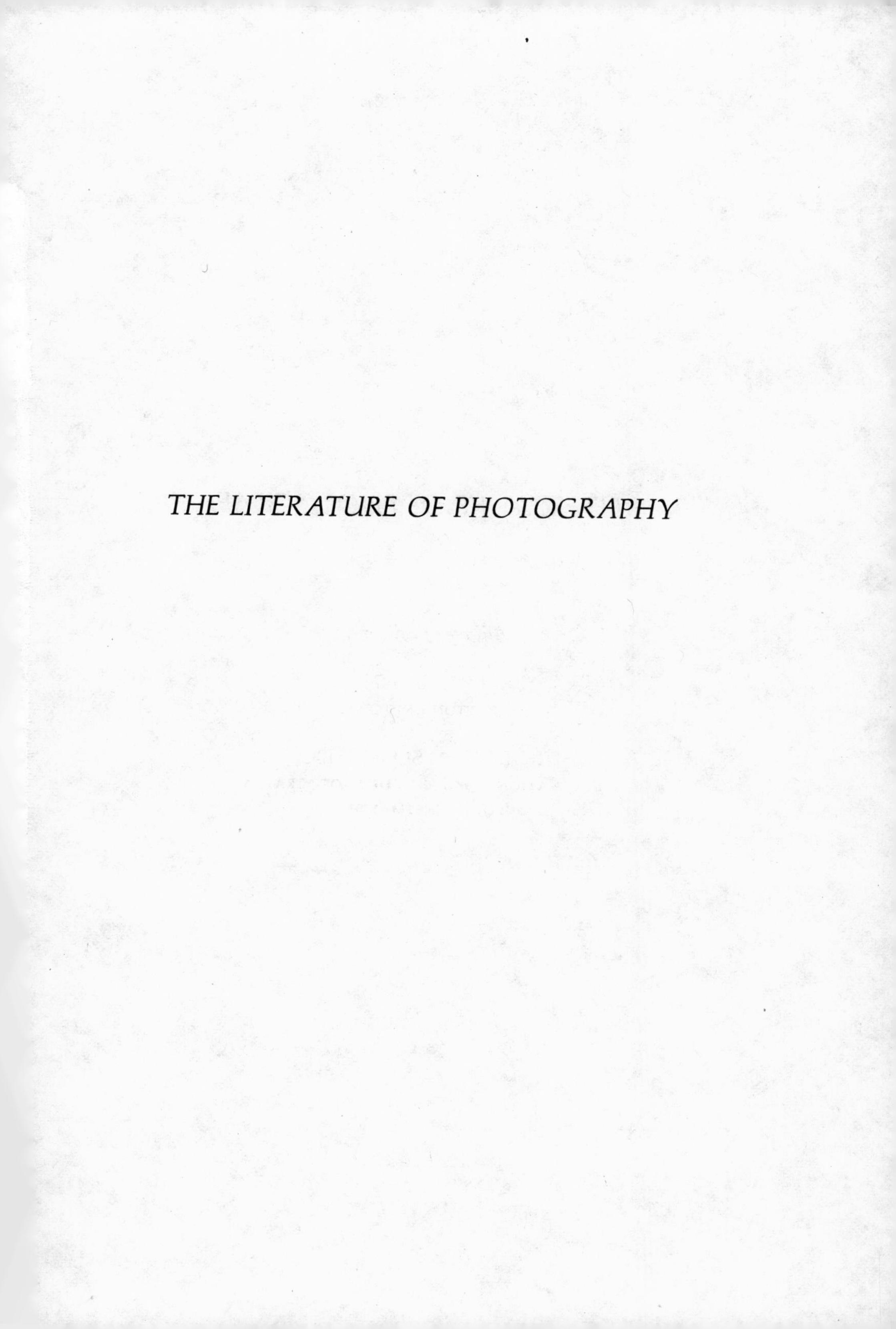

THE LITERATURE OF PHOTOGRAPHY

THE LITERATURE OF PHOTOGRAPHY

ANIMAL LOCOMOTION

THE

MUYBRIDGE WORK

AT THE

UNIVERSITY OF PENNSYLVANIA

ARNO PRESS
A NEW YORK TIMES COMPANY
NEW YORK ★ 1973

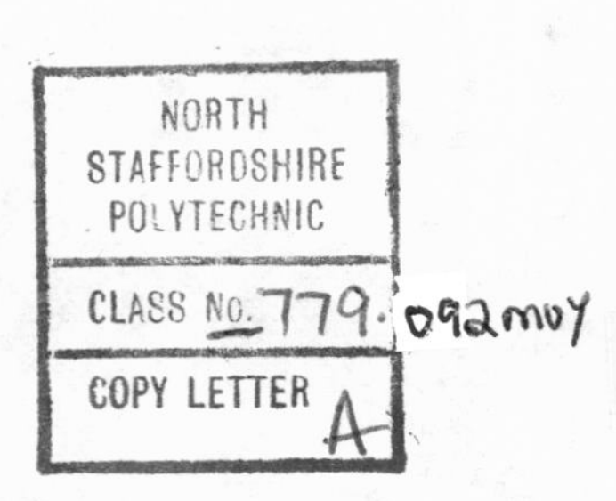

Reprint Edition 1973 by Arno Press Inc.

Reprinted from a copy in
The Harvard College Library

The Literature of Photography
ISBN for complete set: 0-405-04889-0
See last pages of this volume for titles.

Manufactured in the United States of America

Library of Congress Cataloging in Publication Data

Pennsylvania. University.
Animal locomotion.

(The Literature of photography)
Reprint of the 1888 ed.
CONTENTS: Marks, W. D. The mechanism of instantaneous photography.--Allen, H. Materials for a memoir on animal locomotion.--Dercum. F. X. A study of some normal and abnormal movements photographed by Muybridge.
1. Animal locomotion. I. Muybridge, Eadweard, 1830-1904. II. Marks, William Dennis, 1849-1914. III. Allen, Harrison, 1841-1897. IV. Dercum, Francis Xavier, 1856-1931. V. Title. VI. Series.

QP301.P36 1973 591.1'852 72-9239
ISBN 0-405-04944-7

ANIMAL LOCOMOTION.

THE

MUYBRIDGE WORK

AT THE

UNIVERSITY OF PENNSYLVANIA. —

THE METHOD AND THE RESULT.

PRINTED FOR THE UNIVERSITY.

J. B. LIPPINCOTT COMPANY,
PHILADELPHIA.
1888.

J·B·LIPPINCOTT·COMPANY.
DROIT ET AVANT
PRINTERS

CONTENTS.

NOTE.

It is fitting that a few words should be said here in regard to the connection of the University with this work, and the motives which led its authorities to assume the supervision of it.

The function of a university is not limited to the mere instruction of students. Researches and original investigations conducted by the mature scholars composing its faculties are an important part of its work, and in a larger conception of its duty should be included the aid which it can extend to investigators engaged in researches too costly or elaborate to be accomplished by private means.

When ample provision is made in these several directions we shall have the university adequately equipped and prepared to exert fully her great function as a discoverer and teacher of truth.

This book is the result of an unusually elaborate investigation thus conducted during the past four years under the auspices of the University of Pennsylvania. In 1883 it was found that Mr. E. Muybridge, who was the first successfully to apply instantaneous photography to the study of Animal Motion, had long cherished the desire of making a new and greatly extended study of the subject, but had been deterred by the want of means. An extensive photographic plant, elaborate electrical and mechanical apparatus, access to typical animals of many kinds, much space for the erection of laboratories, and a large fund for current expenses were needed.

It was represented to the Trustees of the University that several individuals, appreciating the importance of the proposed work to art and science, would unite in guaranteeing all expenses connected with the investigation if a University Commission should be appointed to supervise the entire affair, and thus insure its thoroughly scientific character.

In March, 1884, the following Commission was appointed for the above purpose:

William Pepper, M.D., LL.D.

Joseph Leidy, M.D., LL.D., Professor of Anatomy.

George F. Barker, M.D., Professor of Physics.

Rush Shippen Huidekoper, M.D., Professor of Veterinary Anatomy and Pathology.

William D. Marks, Whitney Professor of Dynamical Engineering.

Lewis M. Haupt, Professor of Civil Engineering.

Thomas C. Eakins, of the Academy of Fine Arts.

Edward H. Coates, Chairman of the Instruction Committee at Academy of Fine Arts, *Chairman*.

Harrison Allen, M.D., Emeritus Professor of Physiology, *Secretary*.

The investigation proved even more prolonged and costly than had been anticipated, and more than three years elapsed and nearly thirty thousand dollars were expended before its completion.

The result has, however, fully justified the action of the University, as well as the expenditure of time and money; the work will undoubtedly be of lasting service to art and science.

Special mention should be made of the valuable assistance rendered by the authorities of the Zoological Gardens, and particularly by Mr. Arthur E. Brown, the superintendent, who extended every facility to Mr. Muybridge in photographing many typical animals in their splendid collection. It is due to Dr. F. X. Dercum that acknowledgment should be made here of the great amount of time and labor devoted by him to furthering the success of this investigation.

The mass of novel material presented in this work is so great that it has not as yet been possible to subject any considerable portion of it to critical examination. As, however, the sole object which induced the University to assume supervision of this work was to contribute to the scientific study of animal motions, it has been decided to publish in the present form a brief description by Professor W. D. Marks, Whitney Professor of Dynamical Engineering in the University of Pennsylvania, of the apparatus and methods employed; a memoir by Dr. Harrison Allen, Emeritus Professor of Physiology in the University of Pennsylvania, on some of the laws or principles elucidated by Mr. Muybridge's photographs; and an article on the clinical aspects of certain nervous affections as illustrated by instantaneous photography, by Dr. F. X. Dercum, Instructor in Nervous Diseases in the University of Pennsylvania.

WILLIAM PEPPER, *Provost.*

THE MECHANISM

OF

INSTANTANEOUS PHOTOGRAPHY.

BY

WILLIAM DENNIS MARKS, PH.B.C.E.,

WHITNEY PROFESSOR OF DYNAMICAL ENGINEERING UNIVERSITY OF PENNSYLVANIA.

A CAREFUL search through the files of the *British Journal of Photography*, *The Photographic News*, and other analogous papers, reveals such an endless variety of ingenious devices for instantaneous shutters and their appurtenances as will at once prevent even an attempt at a chronological reference to them.

We are forced, by lack of space, to content ourselves by referring those interested to these papers, and to the published writings of Marey, Pickering, and Bowditch.

In the brief description of Professor Eakins's apparatus, hereafter given, we have all that Marey has used, with additions perfecting it greatly.

Professor Pickering has published two interesting papers on instantaneous photography. One is published in the *Proceedings of the American Academy of Science*, January 14, 1885; the other, describing a tuning-fork method of measuring exposures, was published in *Science*, November 14, 1884. It is, however, with the work of Mr. Eadweard Muybridge we shall principally concern ourselves. His work, covering a period of years and involving an almost incredible amount of pertinacious labor, required on his part, and the part of the committee and his assistants, the creation of an elaborate and complicated series of machines, hitherto untried or tried unsuccessfully, and which, furthermore, demanded that skill on their part that can only come from long practice in their use.

No small part of the success of this work is due to the patience

and skill of one of his principal assistants, Mr. Lino F. Rondinella, and it is to him that the writer is indebted for the descriptions, drawings, and diagrams of the apparatus used. Most of the work done in instantaneous photography has been of an unsystematic and desultory nature and limited scope. In this work, which still leaves much to be done, we have the first systematic attempt to record the motions of a variety of animals.

Professor Thomas Eakins, director of the Pennsylvania Academy of the Fine Arts, undertook, under the auspices of the University of Pennsylvania, an investigation of some mechanisms of animal movement, in which he is much interested. His apparatus will be of great interest. He chose as his principal means the photographic method of Professor Marey, of France. The object to be studied is moved in sunlight before a dark background, preferably a deep recess painted black. A photographic camera is set up. A disk with one or more openings in it is rapidly revolved in front of the lens. While the opening is passing the lens the moving object is photographed; then darkness in the camera until an opening again arrives. The moving object has now a new position, and is photographed anew on the same plate, and so on again and again as often as required, or until the object has moved beyond the range of the lens or plate. On the same plate is then developed a series of images easily compared. With the present sensitiveness of the gelatin dry plate a tolerably good image may be had of any light object moving in the sunlight not too fast. If the object moves too fast, the speed at which the opening passes the lens must be increased, or the opening itself narrowed, either of which diminishes the amount of available light, and the image becomes weak.

So in adjusting the instrument for a contemplated fast movement one must choose somewhere between an image weak but sharp and an image strong, but blurred from the movement.

A brief description of the whole apparatus as used by Mr. Eakins may be interesting to the physiologists who may wish to avail themselves of this modern instrument of research.

Instead of one large disk, as in Professor Marey's apparatus, he used two small ones on the same arbor, but geared at different rates of speed,--namely, one to eight.

If a clock-face should have disks instead of hands, and the

openings were made in the disks where the hands should be, and a camera lens placed behind the noon-mark of the clock, an image would be got at twelve o'clock. The fast disk would return its opening in an hour, but the two openings would not coincide till $5\frac{5}{11}$ minutes past one, which would be past the lens.

The two disk openings would again come together at $10\frac{10}{11}$ minutes past two, but the three openings and those in the disks and of the lens would not coincide till twelve o'clock, when another image could be impressed.

The disks were circular saw-blades with four balanced openings, over which openings gossamer waterproof was stuck by Venice turpentine, thickened by heat and spread on the edges. In this waterproof openings were cut radially as wanted.

A crank turned by hand communicated to the disks through a belt going from a large to a small pulley the requisite speed.

Between the disks and the lens was fitted an electric shutter. The disks revolving, and the object set in motion, the observer, with his fingers on the key-board, presses the first button when he wishes the impressions to begin, and the second button when he wishes them to end.

Pressing the first button breaks contact in an electric circuit that holds against the force of a rubber spring, the armature of an electro-magnet controlling the trigger, which releases in the shutter an opaque slide covering the lens opening. The second button of the key-board releases in the same way a similar opaque slide, which, having been held in place above the lens, now comes down in front of it.

The shutter was constructed to move very rapidly. The slides are little squares of silk waterproof, kept stretched by pieces of quill, silk wrapped, attached to the corners. These quills slide vertically, on brass rods, and the slides are pulled down by india-rubber bands, which reach down all the way to the floor. The bands used are the little parcels bands, looped end to end until sufficient length is got.

The reason for using long bands is this: friction overcome, rapidity is got from length, not thickness of spring. For instance, two gum bands, side by side, stretched and released, would not travel faster than one of them; but if looped end to end the speed of the one adds itself to the speed of the other.

A piece of catgut arrests the action of the spring without drawing taut the adjusted strings which pull down the slides.

The closed circuit was used for the shutter, that no time should be lost by the magnet reaching for the armature and because less battery power would be required.

In the camera a new precaution was taken against extraneous light,—a moving diaphragm immediately in front of the plate, and connected with a peep-sight above the camera. An assistant having adjusted the diaphragm somewhat larger than the image to be photographed, and having placed his eye to the hind sight, follows the moving object by keeping it between the two front sights, the front sights being at a distance from the hind sight equal to the conjugate focus of the lens.

The plate itself can also be slid right and left by the frame which carries it, in case successive *poses* may be executed in one place and their images are desired on one plate.

To increase the range of the instrument, a second camera was devised for investigating the class of movement in one place, as that of a man throwing a stone, where, in the Marey camera, the images would superimpose.

In this camera the plate is revolved on a frame at the end of an axle, which pierces the back of the camera.

Belting and pulleys connect the movement of the disks and of the plate, and a pair of friction-wheels allows such an adjustment of speed from the disks to the sensitive plate that the images may be made just to clear one another.

To use this camera the ground glass is set true by adjusting screws and the lens focused. A ten by twelve inch plate or smaller is cut to an octagon and replaces the ground glass, being sprung up against the adjusting screws. The distance of the images apart having been determined upon in degrees when the ground glass was in, the speed of the plate is now adjusted. A style on the armature of an electro-magnet in the exposure-counting circuit dots on a smoked disk, whose rim is graduated, the different exposures as the machine is turned. The sliding-bar carrying one of the friction-wheels is pushed in or out until the dots become the requisite number of degrees apart.

A moving diaphragm is also attached to this camera, having a slight horizontal movement and governed by a peep-sight.

For recording the times and durations of exposure a chronograph was used, which had been designed by the writer for the University of Pennsylvania. An electric current vibrating a tuning-fork in the usual manner, an electro-magnet with a style to its armature in the same current, on the smoked paper on the turning cylinder a series of dots one one-hundredth of a second apart, furnished the regular time-comparison intervals. Alongside of this electro-magnet a similar one recorded the exposures.

The horizontal cylinder on which the paper was stretched and smoked could be turned by hand or clock-work, and the two magnets were on a carriage, which travelled slowly from left to right by an endless screw, as in the slide-rest of an engine lathe.

The means of establishing the circuit which marks the openings or exposures was as follows: An interrupter disk of brass and hard rubber was fixed on the axis of the disks, and is turned on the axle until the brass may correspond to the opening of the disks and there clamped.

A spring with platinum contact point, and capable of movement in two slots at right angles, is also adjusted according to the number of openings and sectors of brass.

The circuit carried through the interrupter is taken to the keyboard that controls the shutter, and the pressing the first key that breaks the contact holding the first slide allows contact in the exposure-marking circuit. The key carries on a spring a platinum point pressing against either of two metal strips according to its position.

The inner straight edges of these metal strips are not parallel, but approach at one end ; and the contact spring being fastened to the key-bar by screws through a slot instead of holes, the spring may be pushed back and forth along the bar, and so adjusted that there is no appreciable interval between leaving the first contact and gaining the second.

Thus the counting begins as soon after the release of the first shutter-slide as an opening comes opposite the lens. These keys have also the ordinary adjusting-screws of the telegraphic key.

The pressing of the second key, which releases the second shutter-slide and stops the lens, breaks the exposure-counting circuit by means of a see-saw between the two keys, which throws the first key back into its first position.

The exposure-counting circuit possesses another device.

On top of the first camera described, where a peep-sight carrying the diaphragm follows the movement, are two parallel strips of metal on wooden slides. They are so slid past one another that the length of metal opposite may equal the length of the photographic plate used. The peep-sight bar carries a bridge clamped at pleasure, springing its contact points from one to the other slide, so that the circuit passing this bridge may not count images that might occur where there is no plate to receive impressions.

The battery used was composed of eight Le Clanché cells. Two, three, or four cells in series are thrown by a switch into the fork, or time-circuit.

Two cells are used for the exposure-counting circuit, and one cell for each shutter-slide, eight in all, and a ninth cell is in reserve to replace or reinforce any one that might accidentally give out. The lens used was a Ross *carte-de-visite* lens of six and three-quarter inches equivalent focus. It is to be regretted that Professor Eakins's admirable work is not yet sufficiently complete for publication as a whole.

FIG. 3.

The reproduction of a boy jumping horizontally, shown above, which Professor Eakins has photographed on a single plate by means of his adaptation of the Marey wheel, is of exceedingly great interest, because, in this picture, each impression occurred at exact intervals. The velocity of motion can be determined, by measurement of the spaces separating the successive figures, with very great precision, as also the relative motions of the various members of the body. The advantages arising from this method of photography would seem to render its further prosecution desirable, as yielding a means of measurement as near scientifically exact and free from sources of error as we can hope to reach.

In 1872, Mr. Eadweard Muybridge, of California, made the first lateral photograph of a horse trotting at full speed, for the purpose of settling a controversy among horsemen as to "whether all the feet of a horse while trotting were entirely clear of the ground" at any one instant of time. It was not until 1877, however, that he conceived the idea that animal locomotion, which was then attracting considerable attention through the experiments of Professor Marey, might be investigated by means of instantaneous photography, with results of probable value to the artist as well as to the student of science and philosopher.

Marey had investigated the action of a horse by fastening to the shoes of the animal elastic air-chambers which were connected by means of flexible tubes to the pencils of a chronographic mechanism carried in the hand of his rider. At each impact of the horse's foot with the ground the compression of air in the chamber caused the pencil connected with it to trace a record on the revolving drum of the chronograph, and in this way much valuable information was obtained regarding the relative action of the four feet of a horse in his different gaits.

Mr. Muybridge proposed to carry this investigation much further by means of photography. His plan consisted in arranging a number of photographic cameras side by side in a line parallel to a track over which an animal should be made to move, so that by means of suitably arranged apparatus an instantaneous photograph should be made in each camera of the series successively, at regulated intervals of time or distance, while the animal was making a complete stride or cycle of movements in front of the battery of cameras. The results would then show the animal in the several successive positions which he assumed in performing any particular movement, and would thus furnish a complete and indisputable demonstration of the disputed problem of animal motions.

Acting upon this plan, a few sets of photographs of horses moving at their various gaits were taken in 1878 with electro-photographic exposors or shutters devised by Mr. Muybridge especially for that work; and although the experiments were made with wet plates in the heat of a California summer, the results were wonderfully fine. They were published the same year by Hon. Leland Standford, under the title of "The Horse in Motion," and (together with others made in 1879) were exhibited by Mr. Muybridge in lectures delivered before the most important scientific and art societies of this country, France, and England. While abroad he met with such hearty appreciation of the value of his pictures, especially on account of the positive facts which they gave in contradiction to many of the accepted theories of animal locomotion, that he decided to pursue the subject still further with as exhaustive a set of photographic investigations as could be made into all the various motions of men, women, and children, animals and birds. Moreover, these new experiments were to be made with new and greatly improved apparatus and methods, with the new "dry plates" that had made possible an amount of detail in instantaneous photography unattainable with wet plates, and with the addition of devices and instruments of accuracy for automatically recording the data of each set of pictures, distance, time, etc., necessary to make them of practical value for scientific purposes. Upon the conclusion of the investigations the photographic results were to be systematically arranged and printed without retouching by photogravure, and finally published by subscription in a manner befitting the magnitude and importance of the work.

After unsuccessful application had been made to several publishers and institutions of learning for the necessary means to carry out these plans, the University of Pennsylvania took the matter under its care, and by its liberal support Mr. Muybridge was enabled to carry on his work in the most thorough manner to its completion, with the gratifying results now shown in his "Animal Locomotion."

All of the photographs for "Animal Locomotion" were made in Philadelphia,—the birds and wild animals at the gardens of the Zoological Society, the thoroughbred horses at the Gentlemen's Driving Park, and the human subjects, carriage- and saddle-horses, and other domestic animals in the studio at the University of Pennsylvania. This out-door studio was built on the grounds surrounding the University Hospital, far enough away from a principal thoroughfare to be undisturbed by passers-by. Its ground-plan, diagrammed in Fig. 1, was somewhat

FIG. 1.

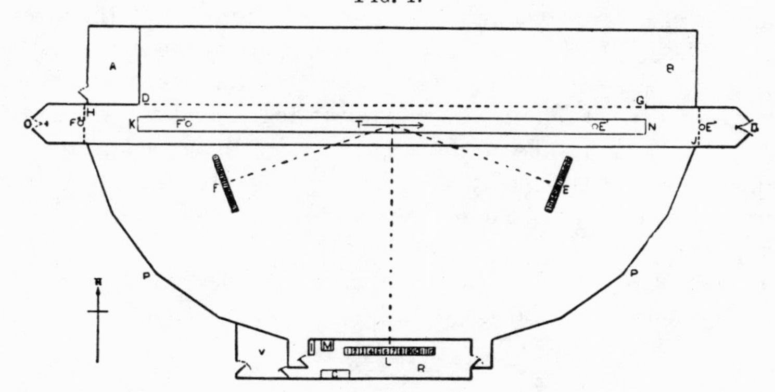

similar in shape to a semicircle. On the north side the diameter of the semicircle was a shed, A, B, one hundred and twenty feet long, eleven feet high, and sixteen feet deep, painted black. The greater portion of its front, D, G, from the edge of the roof to the ground, was covered by a net-work of white threads, accurately vertical and horizontal, and five centimetres (almost two inches) apart, every tenth thread being thicker than the rest. Directly in front of this background ran a level track, H, J, eight feet wide, covered down the centre with heavy corrugated rubber matting, K, N. Each end of the track was closed by a double gate, O, painted black, in front of which hung an adjustable series of vertical threads, H and J, five centimetres apart, every sixth thread being thicker than the rest.

On the opposite or southern side of the enclosure was the camera-house R, a long, low structure with white front, forming, as it were, a chord in the semicircle of the fence P, which extended from each end of it to the ends of the track. The fence was built in this shape and slanting outward at the top, and the inside of it was painted white, like the camera-house, that they might reflect light towards the centre of the track. The entire length of the camera-house front, thirty-two feet, could be thrown open by a number of shutters about two feet wide swinging upward. Inside the camera-house, and running almost the whole length of it, was an immovable table or counter, on whose level top were placed, side by side, a battery of twenty-four four by five inch cameras, twelve of which are represented at L, with their lenses at a perpendicular distance of about fifteen metres (forty-nine feet) from the line of progressive motion on the track indicated by the arrow I, and fifteen centimetres, or six inches, apart between successive centres. These lenses are perfectly rectilinear, three inches in diameter and fifteen inches equivalent focus, and were ordered from England by one of the guarantors of the expenditures assumed by the University of Pennsylvania expressly for these investigations. In front of each camera was an electro-photographic exposor, an improvement on those which Mr. Muybridge had previously invented and used in California. A view of the camera-house, with the row of large exposors inside of it, is shown in Fig. 4.

Fig. 4.

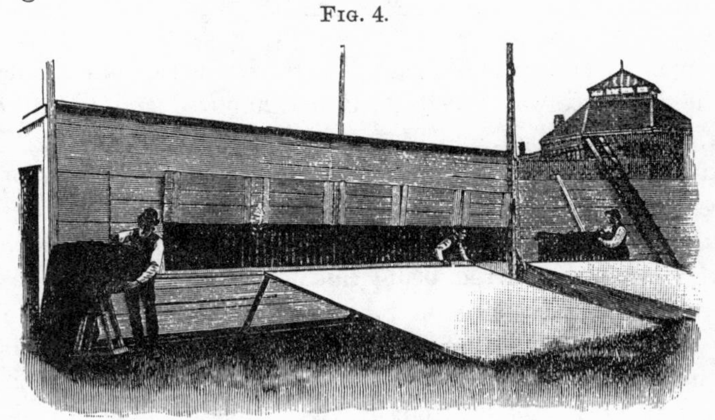

Fig. 5 is a diagrammatic side view of an exposor and its releasing magnet. The exposor proper, A, is a continuous curtain of black rubber-cloth, thin but perfectly opaque, moving easily on the two rollers R and R. In this curtain are two openings, O and O, so placed that they come directly opposite each other in front of the lens L. Motion is imparted to the curtain by the tension of stout rubber bands, B, the duration of the exposure being varied at will by the thickness and number of bands used. Firmly fastened to the front of the curtain is a wooden pin, C, to which are attached a strong tape, D, with a ring in the end of it, and another pin, F, over which the rubber bands B are passed.

The releasing magnet consists of an electro-magnet, M, whose armature is fastened to the lower end of an arm, I, pivoted at H, and having a shoulder on its upper end that may be placed over the extremity of a three-inch steel lever, G, and automatically held there by the pressure of the band-spring S. The lever G is pivoted at a point about three-sixteenths of an inch from its back end, in which is a slight groove or indentation. To set the exposor previous to making a photograph, the front end of G is caught under the shoulder on I, the hold (usually very slight) being regulated by the set-screw N; the ring in D is placed over the back end of G, and the rubber bands B are then stretched to the required tension by pulling upward on the end of a stout cord passing under the pulley P until the ring in the end of the cord can be caught on a hook in the frame of the exposor. A plate is put in the camera and it is ready for action.

Fig. 5.

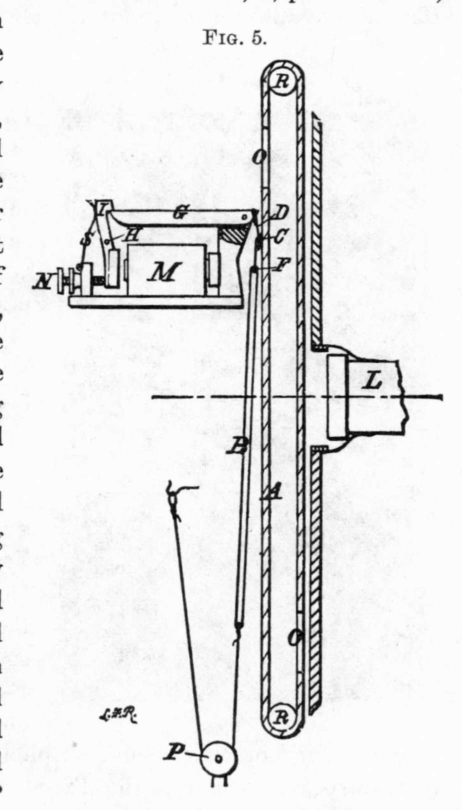

At the proper instant, by means of automatic apparatus to be described farther on, an electric current is sent through the coils of the magnet M; the armature is pulled towards it, thereby releasing the lever G, and the rubber bands, no longer restrained,

fly down to their normal length, carrying the curtain around on its rollers with great rapidity; the openings flash past each other and the impression is made on the plate. Fig. 6 is a front view of three exposors, in the first of which, at the right, the exposure is just ending; in the second it is just beginning, while the third is still set. In this form of exposor, it will be noticed, the opening or exposure begins, reaches its maximum, and ends directly opposite the centre of the lens.

Fig. 6.

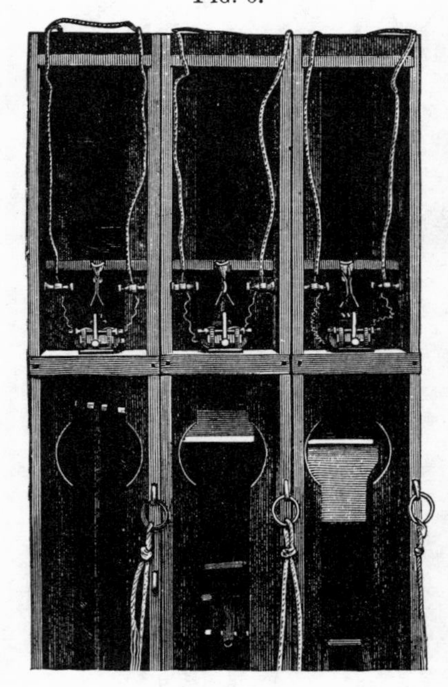

Two releasing magnets are shown in Fig. 7, one of them (A) set, the other (B) released.

The twenty-four large cameras (or twelve, as the case might be), each with its separate exposor, as just described, were used in making the lateral series of photographs of all the motions investigated at the studio excepting the most rapid ones, and with excellent results. But it can be seen that they were not readily portable, and therefore inadequate to do the work at the Zoological Gardens and the Gentlemen's Driving Park in the short time that could be spent at each of those places. Besides this, it was desired to obtain the different phases of a movement not only in *lateral* photographs (those taken from the side perpendicular to the direction of motion, as in the California experiments), but from *any* point of view, or from *several points of view at the same time;* and also to be able to make changes of position quickly upon short notice. These facts prompted Mr. Muybridge to design and have made two batteries of smaller cameras, which more than fulfilled all the requirements of portability and quick working, and rendered possible the making of more than one thousand analytical sets of photographs during the three months spent in his final investigations.

FIG. 7.

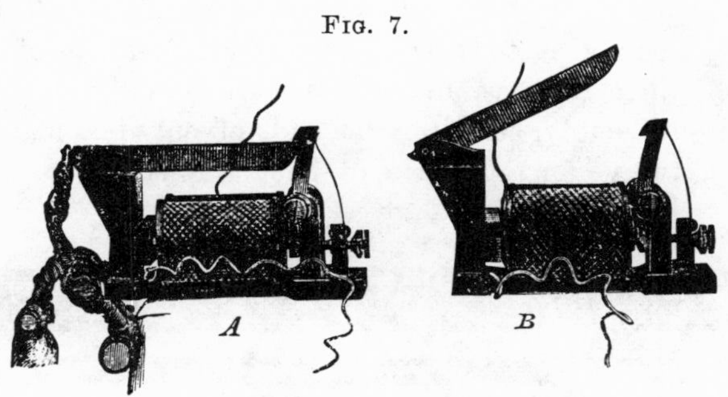

In outward appearance each of these portable batteries (Fig. 8) resembles a rectangular box about eighteen inches square on the ends and four feet long, the back and front being open, and the latter divided by vertical partitions into fifteen spaces, twelve of which are equal. In each of the equal spaces is a releasing magnet and an exposor exactly similar in construction and operation to the one described above, but only one-fourth as large.

FIG. 8.

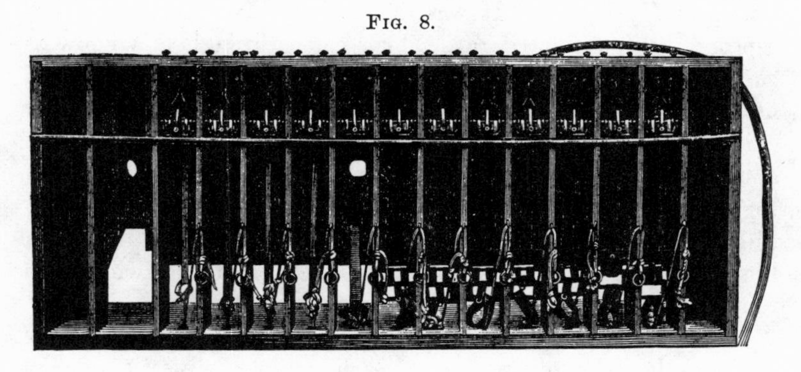

Back of the partitions, and resting upon battens fastened to the side of the box (Fig. 9), containing thirteen cameras in one, for, though each of the thirteen lenses projecting from the front comes from a separate chamber, in construction there is really but one long bellows ingeniously subdivided into thirteen parts. Twelve of the lenses pass into hoods directly back of the double openings in the exposors; the thirteenth opens into one of the extra spaces in the front, and by focusing *it* all of the lenses are focused simultaneously. Three plates, twelve inches long, fill the one plate-holder (Fig. 9) that is used for the whole battery. It can now be understood that there is a great saving of time in the use of these cameras, due to the method of focusing, and to having to fill, empty, and draw the slide of only one plate-holder instead of twelve for each series of photographs.

FIG. 9.

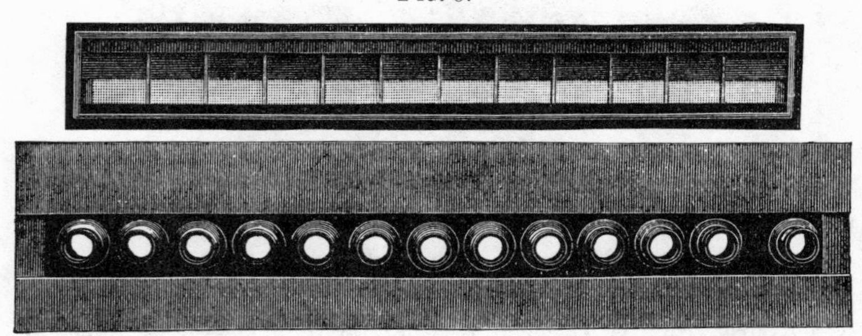

The dark room at the studio in which the plate-holders were filled and emptied was situated at the western end of the camera-house, at V, Fig. 1. It was of necessity absolutely dark, excepting the few colored rays which came through a window of ruby glass covered with dark orange paper; for the dry plates used by Mr. Muybridge were the most rapid he could obtain,—an extra-sensitive emulsion prepared especially for him by Cramer,—and the slightest premature touch of white light would have fogged them. The greatest care was therefore necessary in handling the plate-holders outside. They were carried to and from the dark room each one in a thick black cloth bag, which was removed only under the focusing-cloth of the camera.

Successive exposures were made by sending the electric current through the releasing magnet of each exposor in a series successively, at intervals varying according to the rapidity of the motion photographed. In the California experiments the circuit for each magnet was usually made by the contact of metallic springs, caused by the subject running against and breaking a series of fine threads stretched perpendicularly across the track in front of each camera. But this method, though a very ingenious one, had its disadvantages for the work in Philadelphia. Much valuable time was consumed in arranging the threads for each set of exposures; the time-intervals between successive exposures were unequal and unrecorded; plates were frequently wasted by the subject making the exposures prematurely, or when the character of his motion was unsatisfactory; and, lastly, such a method was impracticable for photographing wild animals and birds.

To overcome these difficulties, Mr. Muybridge used a circuit-breaker which he had devised and unsuccessfully tried in California for making the successive electrical contacts automatically and at equal intervals, long or short, as desired. This was used throughout all the later work. A photograph of this machine, which we shall call the contact-motor, is reproduced in Fig. 10.

Fig. 10.

Two vertical standards support a train of four multiplying gears, to which motion is imparted by a weight hanging from a cord wound on a drum fastened to the lowest shaft, P. Outside of one of the standards, and concentric with the third shaft, is a stationary ring or commutator, composed of twenty-four segments of brass separated by hard-rubber insulation. This is shown more plainly in the two views of the upper part of the apparatus in Fig. 11, and is represented by S in the diagram (Fig. 2), to which future reference letters will apply. To each of the brass segments an insulated copper wire is connected, and passes from thence down through a rubber tube, R, to the bottom of the walnut base of the machine, where it is fastened to the screw of one of the twenty-four binding-posts that are on the base at one side of the apparatus. These can be seen in Fig. 10, and thirteen of them are represented at P in the diagram. Each segment and

Fig. 11.

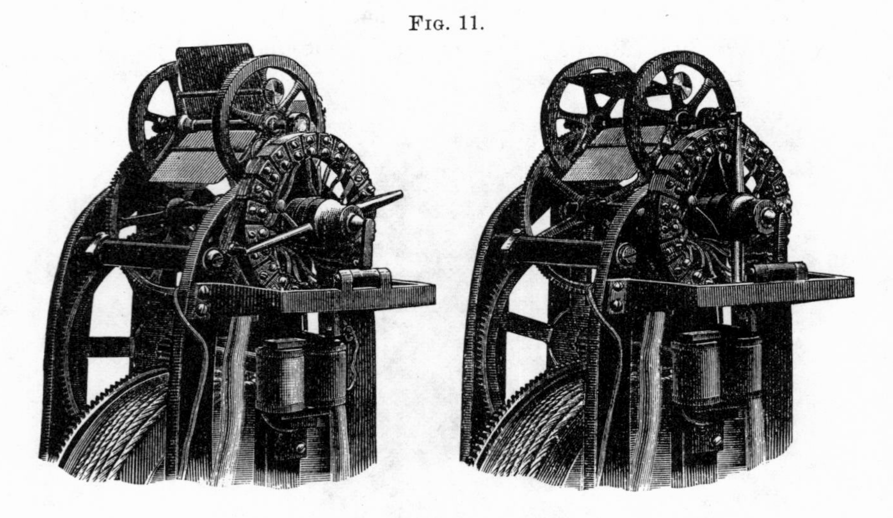

each binding-post is stamped with a number from 1 to 24, and the wire from any segment is connected with the binding-post of the same number.

Loose on the shaft which passes through the centre of the ring is an iron collar carrying the double arm, B, outside the face of the ring. The upper end of B is bent inward at a right angle, and holds a laminated metallic brush in contact with the outside of the segmented ring, while a second metallic brush is held in contact with the iron collar by a stationary pin, K, projecting from the side of the standard. An insulated copper wire connected K with a separate binding-post, U, on the base of the machine. From the lower end of B a small projection or nipple fits into a hole in a lever fastened to the armature of an electro-magnet, M, when the contact-brush on B rests on the insulation between segments 24 and 1. (See Fig. 11.) The armature-lever is held up by a spring until drawn away by the greater force of the electro-magnet, when an electric current is sent through its coils. By an ingenious mechanical arrangement, to which we shall refer more particularly hereafter, when the armature is drawn the loose collar is thrown into gear with a second collar fast to the shaft; and if the shaft is in motion the brush is thereby carried around over the periphery of the ring S, making contact with each of the brass segments successively. The position of the contact-motor in the camera-house is shown at M, Fig. 1.

An insulated copper wire leads from each of the twenty-four binding-posts, P, to the releasing magnet of the large exposor, L, of the same number, while a second wire passes from each of the first twelve of them (P 1 to 12) to a double binding-post, D, stamped with the same number. Two cables, G and H, each over one hundred feet long and composed of thirteen wires, connect at one end to the double binding-posts, and at the other end twelve of the wires to the releasing magnet of each series of portable exposors, F and F, the thirteenth serving as the return wire to D, R, from each series as shown in the diagram. Another wire from D, R, carries the return current from both series of small exposors to a separate binding-post, P R, to which also a wire with twenty-four branches conducts the return current from the large series of exposors.

With the connections made, as just described, a portable battery of cameras could be moved at a moment's notice, as was often necessary, from one part of the field to any other within a radius of one hundred feet of the double binding-posts; and at the Zoological Gardens, on some few occasions, when only one series of exposures was desired, the cables were connected end to end, and then with the operating apparatus at one station the cameras could be moved in any direction to places two hundred feet distant.

The electrical battery, C, which furnished the motive-power for operating the magnets, consisted of fifty-four Le Clanché (prism) cells, usually arranged in a multiple arc of three series of eighteen cells. At the studio they were placed on shelves in the camera-house at I in Fig 1. A wire from the negative pole of the battery carried the current to the binding-post U, while a wire from P, R, carried it back to the positive pole. In the circuit of this return wire the electro-magnet *e, m,* of the chronograph was placed, and by the arrangement a style fastened in its armature traced a record on the lampblacked paper covering the drum J

FIG. 2.

ANIMAL LOCOMOTION.

DIAGRAM OF ELECTRICAL CONNECTIONS FOR MAKING CONSECUTIVE PHOTOGRAPHIC EXPOSURES SYNCHRONOUSLY FROM SEVERAL POINTS OF VIEW

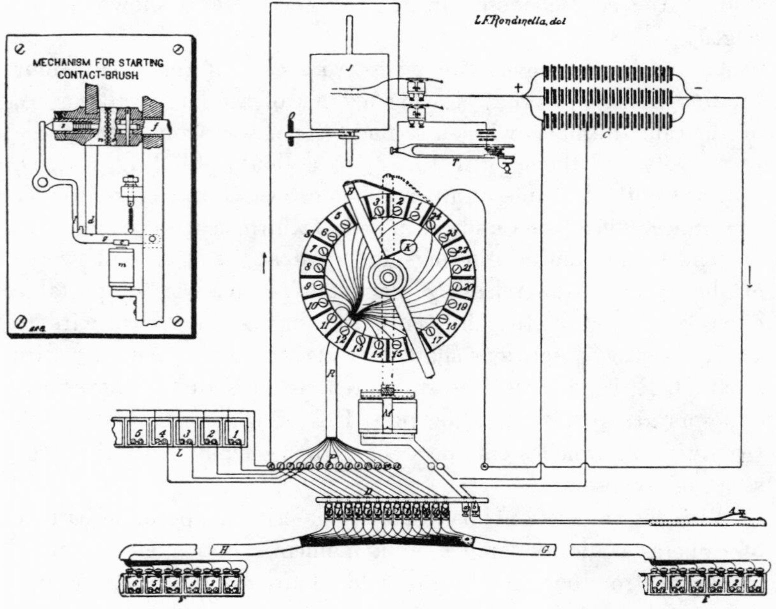

whenever the armature was drawn down by the passage of the current,—*simultaneously with the commencement of each exposure* in a series of twelve or twenty-four. While this record was being made a time record was made directly under it by a style from the electro-magnet *t, m,* in circuit with a tuning-fork, T. The tuning-fork used throughout the investigations was pitched to one hundred single vibrations per second, the vibration being maintained by a separate electric battery of two cells. The drum J was rotated on its vertical axis by clock-work, and could be moved up or down on the shaft by means of a supporting arm working on a screw, O, so that the paper could be entirely covered with records (often fifteen or twenty) before renewing. The drum and shaft were then lifted out of the bearings, the paper cut off, and the records fixed permanently upon it by immersion in dilute shellac. The chronograph, its batteries, etc., were enclosed in a wooden box, which at the studio stood on a shelf at C, Fig. 1. The electro-magnet M, whose purpose has been already explained, was operated by a shunt current from six cells of the battery C, the circuit being closed by the contact-key A. The latter was designed especially for portability, and consisted of two stiff brass springs with platinum contacts, mounted on round sticks about two feet long, so that it could be conveniently carried in the hand from place to place; for the operator usually stood directly beside the camera in which the first photograph of the series was to be made, and by making the contact when the subject reached the desired position the whole series of exposures was started.

Having now described the apparatus in detail, we can, perhaps, best understand the method of operation by imagining ourselves in the studio at the University, watching the making of a series of twelve photographs from each of three points of view simultaneously, thirty-six different negatives in all. In order to make our observations more definite we will suppose that the action to be investigated is one stride of a walking horse; that the large or "lateral" cameras, L, Fig. 1, are in their usual position parallel to the line of motion on the track and at a distance of forty-nine feet from it; that the portable cameras, E and F, have been placed, one battery for front the other for "rear sixty-degree foreshortenings," at a distance of thirty-five feet from the centre of the track; that is, in places permanently marked so that they point in a direction at nearly sixty degrees to that of the laterals on both sides; the lenses of all three batteries are in the same horizontal plane with a certain heavy thread in the lateral background, the "horizon line."

The cameras being already focused, an assistant proceeds to set the exposors, and the horse is started on a trial-trip down the track that we may see about how much time he occupies in making one *s* stride. We count ten strides in eleven seconds (those of us who are experienced in such things may, perhaps, remark that he is a rather slow walker), and another assistant then adjusts the contact-motor by altering the weights and fan governor so that the contact-brush will pass over the first twelve segments of the ring in about one and one-quarter seconds, making a slight allowance for safety; "because," the assistant will tell us, "we would rather get the whole stride in only eleven phases than run the risk *of not* getting it in twelve, if the horse should decide to saunter a little." Next he gets the chronograph ready for action, and the other assistant meantime having finished setting exposors, brings the necessary plate-holders from the dark room—twelve small ones for the lateral cameras and one large one for each portable battery—and places them in the cameras. The slides are drawn and we are all ready.

The operator, with the contact-key in his hand, takes his stand beside camera No. 1 of the lateral series, and his assistant inside the camera-house starts the contact-motor and the chronograph while the horse comes walking down the track. When he reaches a point directly in front of camera No. 1 the operator presses the contact-key; we hear a slight humming sound, and thirty-six different photographs of the horse are made, three different views of each twelve successive positions that he put himself into while making a single stride. If now we step inside of the camera-house we can see on the drum of the chronograph a record that was made simultaneously with the exposure; and if we are curious to know we can count the number of vibrations in the lower line of the record lying between the first and twelfth drop in the upper line, and thus see exactly how many hundredths of a second were spent in making the series of exposures. Turning again to the diagram, Fig. 2, we will see how a momentary contact of the key brought about these results. The drum of the chronograph, the tuning-fork, and the shaft through the collar of B are all in motion, while B is held in the position indicated by the dotted lines by a catch in the armature-lever of M, as before described. In this position the brush rests on the insulation between segments 24 and 1, and therefore no circuit for the electric current. Now contact is made for an instant at the key A, a current passes through M, its momentary magnetic attraction pulls the armature-lever down, thereby freeing the arm B and at the same time throwing it into gear with the rotating shaft. The brush is thus carried around over the ring S, and as it makes contact with each segment, a current passes through the three exposor magnets of the same number and through the chronograph magnet, and exposures are made synchronously from the three points of view. With the brush in the positions shown by full lines in the diagram we can follow the course of the current; from the negative pole of the battery C to the binding-post U, thence to the stationary brush K in contact with B, through B and the rotating brush to segment No. 2, from there through R to binding-post P 2, where it separates into two parts, one passing through exposor magnet L 2, and back to P, R, the other to double binding-post D 2, where it subdivides into two parts, passing through exposor magnets E 2, and F 2, respectively, and back through D, R, to P, R; from P, R, the entire current returns to the positive pole of the battery C, after passing through the chronograph magnet *e, m*, and thereby making a record of that exposure on the drum J.

The way in which the contact-brush is automatically thrown into gear with the revolving shaft can be seen by reference to the diagram of "mechanism for starting contact-brush" (Fig. 2), where some of the revolving parts are drawn in vertical section to show the internal construction. The shaft *f* is bored hollow for a part of its length, the internal diameter being smaller near the bottom of the bore than it is towards the end of the shaft. Into the bore fits a spindle, S, whose external diameters correspond to the internal diameters of the shaft. To the spindle S is keyed a collar, O, in whose outer face radial teeth are cut. The key V passes through a slot in the shaft F, so that the spindle S, with the collar O, can be moved in or out for a short distance, limited by the length of the slot, while at the same time it will be noticed S and O will be carried around with *f* when it revolves. Loose upon *f* is a second collar, *n*, which carries the double arm *d* with the contact-brush, and has on its inner face radial teeth to gear into those in *o*. In the space inside the shaft, between the shoulders in *f* and that in *s*, a coil spring is fitted, which pushes the spindle outward, throwing *o* into gear with *n*, unless prevented by resistance at the conical extremity of *s*. Such resistance is offered by the upper arm of a bent lever, *l*, whose lower arm fits in a notch in the extremity of the armature-lever *g*, the latter being held in place by a spring. The double arm *d* also catches in *g*, as explained before. Now, with *g*, *l*, and *d* thus arranged, as shown in the diagram, let us suppose *f*, and with it *s* and *o*, to be revolving at the desired speed. An electric current is sent for an instant through the coils of the electro-magnet *m*. Its consequent magnetic attraction pulls down the armature *c*, and with it the lever *g*. The spindle *s*, no longer opposed by *l*, is pushed outward by the force of the spring; *o* is thus thrown into gear with *n*, and the contact-brush on the arm *d*, now free to turn, is carried around with the revolving shaft. When the armature-lever *g* is pulled down as described, the lower end of *l* catches in a second notch in its extremity and holds it down until the motor is reset by hand. The two different positions of these levers can be seen in the two views of Fig. 11.

The most usual positions in which two portable series of cameras were placed when used at the studio for making front and rear views, or "foreshortenings," in conjunction with the first twelve of the permanent series for side views, or "laterals," are shown in Fig. 1, at *f* and *e*, thirty-five feet, at E′ and F′, forty feet, or at E″ and F″, sixty feet from the centre of the track. In the last two sets of positions the cameras pointed directly up or down the track, at right angles to the laterals. Photographs made in them were therefore termed "ninety-degree foreshortenings." Other combinations of position were made by placing one of the portable batteries of cameras at sixty degrees front or rear, with the other at ninety degrees rear or front respectively. In the ninety-degree foreshortenings, which were made especially to analyze side oscillations of the body or limbs during locomotion, it was desirable that the successive points of view in the series should be all in the same vertical plane, and in order to have them so the camera-box was stood on end, with the lenses directly above one another. When the cameras were so placed the "horizon line" was on a level with a point midway between lenses 6 and 7, and the first exposure of the series was made through the lowest lens, except in a few special cases.

The portable backgrounds used in all the photographing away from the studio (except with a few of the wild animals at the Zoological Gardens) were frames four metres long by three metres wide (approximately thirteen feet by ten feet), covered, some with black, others with white, cloth, over which were stretched threads contrastingly white or black, forming squares exactly like those in the lateral background at the studio. They could be set up and levelled in a few moments back of the course over which the animal or bird was to pass, and were firmly held in position by guys.

The background being arranged at the studio or away from it in the manner described for each, it will be readily understood that every photograph in a multiple series had upon it, behind the figure, a number of equidistant parallel lines, both horizontal and vertical in all except the ninety-degree foreshortenings made at the studio, where they were vertical only. Now, by noting the positions of any part of the body upon the background of squares in the consecutive photographs of a lateral series the amount of forward and upward motion between successive phases can be accurately determined; and in the same way, by means of the background in the ninety-degree foreshortenings, the amount of a sidewise movement can be found. From these determinations the curves which different parts of the body describe may be readily plotted. These trajectories, or those described by the same limbs of different animals in performing the same movements and so forth, can be critically compared, and at the same time the corresponding amount of muscular action shown in the photographs can be examined, with results of probably greater interest and value than we can at present foretell.

During the investigations in the summer of 1885 a very interesting double series of photographs was made to determine the equality of the intervals between successive exposures, the accuracy of the chronographic record of these intervals, and the duration of the shortest exposures used in the work, Mr. Muybridge's statement that in some cases the exposure lasted but one five-thousandth of a second having been doubted by several critics. Short photographic exposures are usually measured by photographing some object moving rapidly at a known or readily-determined speed. The blur in the photograph shows how far the object moved while the picture was being taken, and from these data of distance and speed the duration of the exposure is calculated. The usual objects photographed (a falling ball, a swinging second pendulum, etc.) being too slow for such short exposures, a new method was devised, in which the moving body was a circular black disk over five feet in diameter, turned by a crank with a multiplying gear of ten to one, so that for a single turn of the crank there were ten revolutions of the disk, and at high speed the chances of irregularities in the motion were therefore reduced to a minimum. A cam on the axle of the disk made an electrical contact between two metallic springs at each revolution, by means of which the speed of each turn was recorded on the drum of the chronograph. On the black surface of the disk, near its periphery, was a white spot of a certain width; its distance from the centre, and (from that and the chronograph record) the distance which it passed over in the time of one revolution, were accurately known. From the increased width or blur of the spot shown in each photograph the duration of that exposure was calculated; and by noticing the difference in the position of the spot in any two consecutive photographs of the series the interval between successive exposures was determined, it was seen whether they were all equal as they should be, and whether the intervals determined in that way agreed with those recorded on the chronograph by the style in circuit with the contact-motor.

In making the experiment the two batteries of small cameras were placed side by side opposite the disk; they were arranged for the most rapid exposures ever used in the investigations, and the contact-motor was adjusted so that one battery should be discharged after the other, a series of twenty-four consecutive exposures in about one second. When all was ready the contact-motor and the chronograph were started, and the disk was revolved at a speed of about two revolutions per second; the switch in the chronographic circuit of the disk was thrown in, and simultaneously the contact-key was pressed, and the series of exposures were made. A portion of the very interesting chronographic record of the series is reproduced in Fig. 12, the upper line recording the revolutions of the disk, the middle one in the vibrations of the tuning-fork, and the lower one the intervals between exposures. The vibrations in the upper and lower lines, after the drop caused by the passage of the current, were due to the necessarily long styles in those two electro-magnets, and are not so noticeable in the regular two-line records of the other work.

After the series was finished, with the same short exposure two single photographs were made: one of the disk at rest, in order to have the size of the white spot for comparison, and another of a

Fig. 12.

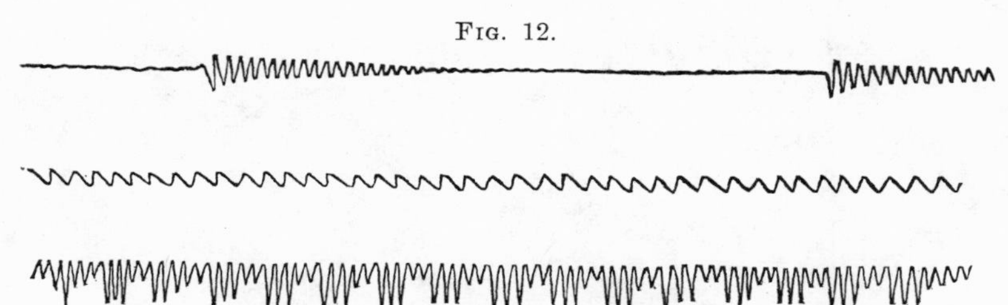

man ascending stairs, in order to show the amount of detail obtained with such short exposures. Subsequently all of these results were carefully examined and worked up by several professors of the University, whose reports stated: 1st, that the intervals between successive exposures were equal except in two cases, where they differed by a few ten-thousandths of a second; 2d, that the intervals between exposures shown by the photographs and those shown by the chronograph were alike, or, in other words, that the chronograph accurately recorded the intervals between exposures; and, 3d, that the duration of the exposure was less than one four-thousandth and more than one six-thousandth of a second.

Upon this third point the writer cannot wholly agree, but has no reason to doubt that the *effective* action of light upon the sensitive film was other than as stated by these gentlemen.

Such very short exposures, however, were only necessary in photographing small objects moving very swiftly, such as birds flying, but the results obtained with them serve to show how great are the possibilities of the application of instantaneous photography in the future.

MATERIALS FOR A MEMOIR

ON

ANIMAL LOCOMOTION.

BY

HARRISON ALLEN, M.D.,

EMERITUS PROFESSOR OF PHYSIOLOGY IN THE UNIVERSITY OF PENNSYLVANIA.

INTRODUCTION.

THE writer has undertaken, at the request of the representatives of the University of Pennsylvania, a series of studies on animal locomotion. The subject has been approached from the point of view presented by instantaneous photography, and has been especially based on the results obtained by Mr. Eadweard Muybridge. He will also state that he is indebted to Professor Thomas Eakins for facilities afforded in studying the results of an experiment in the use of a modified form of Marey's wheel devised by him, in photographing the action of the horse in motion.

Many of the statements could have been deduced from data already accessible to the writer, but since he wrote the paper immediately after the inspection of the photographs his conclusions may be said to be based upon them. At the same time he has not hesitated to include materials not embraced by the photographs. Whenever practicable the study of a given series was carried on at the same time that the animal itself lay dissected before him. For example, when studying the photographs of the elephant, he had the good fortune, through the courtesy of Professor Huidekoper, to dissect the limbs of an elephant. He has also dissected the horse, the ox, the raccoon, the sloth, the skunk, the Virginian deer, and the domestic cat. In the course of the investigation Prof. Horace Jayne extended many facilities. Mr. Edwin A. Kelley has kindly furnished a carefully prepared list of the names of the quadrupeds and birds; this appears in an appendix to the report. Mr. Kelley also gave valued assistance in the dissections.

To better inform himself on the subject of the gait of the horse, the author has observed the impressions made by the hoofs on a sand-beach. In this connection he desires to express his acknowledgments to Dr. M. H. Cryer for valuable assistance and suggestions.

The great delay that would be occasioned in the preparation of this essay by subjecting every statement based upon observation to the test of anatomical demonstration, caused the writer to remain content with making many of the conclusions rest upon the evidence presented by the phenomena recorded by the photographs. He is aware that such conclusions are not in all respects satisfactory. The several heads under which the appearances are discussed must be looked upon as themes which suggest research rather than summaries of work already done. It is hoped that the notes may stimulate some of his readers to attempt to confirm the writer's impressions and further to extend the application which can be made of the pictures to the important subject of animal locomotion.

The "materials" are so arranged as to make two main groups of notes,—those pertaining to the action of the limb when on the ground and those pertaining to the action of the limb about to leave the ground and when in the air. To these will be appended a short section on the human subject, on zoological considerations, and on flight. Occasional repetitions of statements of facts have been difficult to avoid in preparing the notes. The inconvenience arising from this defect has been in a measure overcome by the use of cross-references.

The Use of Terms.

It is necessary to propose the use of a few terms which will clearly express in a word a meaning which otherwise would require the employment of a cumbersome phrase. The words flexion, extension, abduction, adduction, pronation, and supination imperfectly express many animal movements. Thus no adequate word exists for the movement of the hand upward and downward when the forearm is held in semipronation. Such a motion is of importance in describing the wing of the bird and the bat. The term *Planation* is here thought to be convenient, since it expresses the fact that the movement of the hand is in the plane of the flexor surface of the forearm. Planation includes both pronation and supination. *Contraplanation* as easily indicates a movement at an angle to the plane of the flexor surface of the forearm. It embraces flexion and extension, and is not especially demanded, except that a term which includes both flexion and extension as well as abduction, as used in the sense of abduction of the thumb, may be acceptable.

In like manner it has been found useful to employ terms for the different orders of the foot-falls.

Synchiry indicates that the right and left foot of a single pair act together. Thus in the canter, gallop, and the run, the horse moves all the feet synchirally. The movement of the lower limbs in man is also synchiral. The feet may act alternately or together.

In the strictest sense all mammalian limb-motions are synchiral, —*i.e.*, the right and left parts of a single pair move alternately. This motion is at its maximum in the trot, and at its minimum in the spring as performed by the kangaroo. The term is intended to apply to the act in which the feet succeed each other on the ground. The alternation of the right and left hind legs on the ground, while the fore legs are in the air, is a posterior synchiry; and in the same way the alternation of the fore legs on the ground, the hind legs being in the air, is an anterior synchiry. The rabbit appears to resort to synchiry as an alternation of the walk even in the slowest rate of movement; but, as a rule, it may be said that synchiry is an expression of a fast gait.

Asynchiry would naturally embrace the movements in which the feet act in combinations of hind and fore feet. But since these movements are varied and important, it has been thought desirable to substitute a positive term, and the word *heterochiry** is proposed. The walk, the trot, and the rack are heterochiral, since the fore foot is followed, not by its fellow, but by a hind foot. When the

* While assuming the responsibility for this word the writer desires, at the same time, to say that it was suggested to him in a conversation with his friend, Professor T. N. Gill.

hind foot alternates with the fore foot of the same side *lateral heterochiry* occurs. When with the fore foot of one side the hind foot of the opposite side alternates, *diagonal heterochiry* takes place.

In connection with the terms flexion and extension the following will be used: The movement of a limb against the medium in which the animal is moving constitutes the *stroke.* The movement in preparation of the stroke constitutes the *recover.* In the description of the "hand-over-hand" movements of the sloth and the ape, the word "flexion" has no place; yet the "recover" is used in as exact a sense as in the movements of any other animal.

When a limb of a terrestrial quadruped rests upon the ground it may be said to be "on," and when not on the ground, to be "off."

The term *sura* will be employed as a convenient word to include the hind limb from the knee to the ankle. *Crus* has been retained so generally as a synonyme for the entire posterior extremity as not to be available.

"Stroke" is the period of impact. It is included in flexion, and constitutes its first stage. "Recover" embraces the last stage of flexion and the whole of the period of "extension." The terms "stroke" and "recover" are by no means the same as flexion and extension. They simply express certain phases of limb-function which are seen during the acts of flexion and extension.

The support of the trunk begins in advance of the neck—viz., at or in front of the occiput (see p. 54)—and passes back to a point midway between the withers and the rump. In the camel the fore foot supports the trunk at a point as far back as the middle of the posterior dorsal hump. In the same animal the posterior extremity supports the body from the dorsal concavity backward. (See p. 52.)

In the Canadian deer (series 695, Fig. 1) and in the fallow deer (series 682 B, Figs. 2 and 3) the fore limb at one stage of the gait is directly under the trunk, and is the only support of the body. In Procyon (series 744, Fig. 3; see also p. 72) two limbs (the right fore limb and the left hind limb) are in the same position as the foregoing. In the sloth (series 750, Figs. 1, 6, 7, 8, 9, and 12; see also p. 78) the two lateral limbs are also the same.

The right hind foot of the horse (series 631, in the interval between 11 and 12) is in the same attitude. (See p. 43.)

Wherever occurring the position illustrates the tendency for one or two feet of the same, or opposed sides, to support the trunk at some point between the vertical lines of the limbs. It is proposed to call such support *central.* Central support is opposed to the support of the limbs in the extreme or forward motion of the fore limb and the backward motion of the hind limb, which may be called *terminal.*

The Movements of Limbs.

If a limb can be conceived moving in vacuo it can be at once understood that propulsion is impossible. For propulsion can follow only upon the initiation of an impetus, and this in turn only by the resistance of the limb against the medium in which the animal is moving, or, in the case of the terrestrial animal, the surface of the ground.

The resistance of the air and the water is so much less than that of the earth that the acts of flying and of swimming become radically different from those of walking, of running, or of any allied movement. In flying and swimming the resistance made by the limb against the medium in effecting an impetus does not arrest the movement of the pinion or the foot; whereas in terrestrial movements the instant that the foot strikes the earth the resistance is great and the arrest is complete.

In the swimming-turtle the first stage of the recover drives the foot in spite of the resistance of the water to the point at which the second stage begins. With some slight modifications the same is true of fossorial animals. Thus in flying, in swimming, and in burrowing the limb describes a continuous movement which unites the path of the stroke to that of the recover. In the animal moving on the surface of the ground, the foot being brought to rest, an absolute break occurs between the beginning of the act of recover and its completion,—the time which would be required to describe the interval and thus to complete the union corresponds to the period that the foot is on the ground.

The limb rests on the ground until the trunk moves beyond the point at which it can maintain itself. It is lifted at intervals which are dependent upon the momentum of the moving mass.

One, two, or three limbs may be on the ground at the same time. The rates at which the succession of the foot-falls occur, in their turn, depend not only upon the rate of speed at which the animal is moving, but on the gait as well. (See p. 59 and *infra.*)

The Positions of Limbs.

In studying the motions of the limb of a vertebrate the position which answers to that taken by the salamander, when at rest, is assumed to be the best adapted for comparison. In this position the limb is horizontal to the plane of the longitudinal axis of the body. The ventre of the body and the ventral surface of the limb are on the same plane nearly. The limb of a reptile varies scarcely at all from that just named. When a terrestrial animal is erect the limb instead of being on the same plane with that of the body is moved a quarter of a circle downward. In the bird the posterior extremity when at rest is in the same position as the terrestrial, but the anterior extremity, in marked contrast to it, is flexed. When extended the extremity is thrown upward to a position as far removed from the horizontal position of the salamander in one direction as is that of the terrestrial quadruped in the other.

In the movement of all limbs the directions in the main are forward and backward. The forward movement will be described as *forward strain*, and the backward movement as *backward strain.* Both the movements are oblique, but between them is a position which is straight. In the terrestrial animal this position may be said to answer to a line in the anterior extremity which lies immediately in advance of the withers and in the posterior extremity to the centre of the acetabulum.

THE ACTION OF THE LIMB WHEN ON THE GROUND.

Kinds of Work done by the Limbs.

The kinds of work done by the limbs are two in number,—viz., that done by the fore limbs and that done by the hind limbs. The hind limbs are more powerful than the fore limbs, and in some animals, as the kangaroo and the jumping-mouse, are the main effectives. No terrestrial animal depends for support upon the fore limbs. When all the limbs are equal, or nearly equal, in length, the preponderance is still in favor of the hind limbs, owing to the fact that the great backward movement of these limbs on the trunk is made possible by the fixation of the bones to the pelvis, and through this structure to the vertebral column. Not only is this the case, but the hind limbs alone possess the power of propelling the body, so as to throw upon the fore limbs the labor of accommodating themselves to the rate of work of their more powerful associates. When an animal is moving at a high rate of speed, as in the gallop, the synchiral action of the hind limbs projects the body with such force as to compel the fore limbs to act simply as props, which successively carry the body forward until one of the hind limbs is again in position to give the body a second impetus. In proof of this assertion it is only necessary to observe that the greatest height attained by the trunk is that secured by the rump when both hind feet are off the ground. The statement generally made that the horse leaves the ground by one of the fore feet creates the impression that he gains the springing force from this foot, all the previous movements being in preparation for such a spring. In place of this statement another is here substituted,—viz., that the horse springs from that hind foot which last leaves the ground, and is "off" from all feet when he simply relinquishes the support afforded by the last prop,—that is to say, the last fore foot. (See pp. 53, 62.)

If the fore and hind limbs were based on the same plan, the motion of an animal would be either a series of springs—the two feet pushing against the ground at the same moment—or a series of steps, the two feet moving alternately. While closely resembling one another, the two limbs are not on the same plan. If any motion takes place in the vertebral column at the time that the fore limb is moving, it is noticed that it occurs in the region of the neck. The scapula has a slight motion downward and backward. The trunkal motion for the hind limbs occurs in the region of the lumbar vertebræ, while the pelvic bones are fixed. The limit of the forward motion of the hind limbs is dependent upon the flexibility of the lumbar vertebræ. The limit of the similar motion of the fore limb is determined by the action of the muscles alone. The forward motion of the fore limbs is essentially the same in all animals, but the forward movement of the hind limbs is variable, because the lumbar vertebræ differ in degrees of flexibility. In clawed forms there is more lumbar flexibility than in the hoofed. In backward movements the opposite obtains, for in these positions the fore limbs can be carried back to a variable distance. In the deer and its congeners, owing to the great obliquity of the humerus, the fore foot can be brought to a point near the centre of the body, and the limb be vertical. (See p. 92.) In the horse the fore limb in backward strain is very oblique, and the foot, while well placed under the trunk, cannot reach the centre. In the baboon the fore foot cannot pass beyond a vertical line which intersects the trunk a little back of the shoulder-joint. The backward movement of the hind limb is nearly the same in all animals. The leg is always carried in a direct path, the limit of the movement being determined solely by the length of the limb. In a word, the forward movement is the less constrained in the fore limb, while the backward movement is the least constrained in the hind limb. The most variable movements are the backward for the fore limb and the forward for the hind limb.

The "stroke," or period of impact against the ground, can be divided into three stages,—first, that of forward movement, when the foot rests against the ground, chiefly on the outer border (see p. 50); second, the vertical position, when the weight is borne upon the foot as a whole, or when the marginal toes are shorter than the third and fourth upon the central portion; and, third, the backward movement, or backward strain, when the foot rests chiefly upon the inner margin. I am informed by Dr. M. H. Cryer that skaters invariably bring the skate "on" the ice by the outer edge of the runner, and take it "off" by the inner edge. Practically the same motion is seen in swimming for both hands and feet.

The fore leg may be said to be thrust forward as the hand is thrust when used for grasping. The ground may be said to be seized by the foot and the body pulled up to the point of purchase. (See p. 52.) The hind leg (usually of the opposite side) pushes the hind part of the body forward at the same moment. As the fore foot passes its vertical it also would tend to push the body, and in this way assist the hind leg.

In clawed animals the seizing power is carried to a high state, and in hoofed animals it remains at a low one. In all domesticated animals the pushing power of the hind foot is unimpaired. (See p. 92.) It is most modified in forms adapted for tree life, as the ape and the sloth.

Motions essentially the same in all Quadrupeds.

The motions of all quadruped animals are essentially the same. If a comparison is made between the dog and the raccoon, the contrast between the two forms is at first sight great. But if a careful analysis of the figures be made, it will at once be noticed that the points of variance are in the main due to the rate at which the animal is moving, the gait being the same. The flat, broad, yet exceedingly flexible foot of the raccoon is raised deliberately from the heel to the toes, and at each stage of the strain, which is coincident with the heel elevation, the graphic expression of the pictures is totally unlike the short, decisive motion of the small and almost digitigrade foot as seen in the dog. (See p. 50.) But the order of the foot-falls is similar in the two animals.

In like manner contrast may be established between the appearance of the foot as it is being carried through the air in preparation for a stroke. When an animal is going at a high rate of speed the succession of the movements of the fore and the hind feet is altered. In slow rates the fore foot remains on the ground until the hind foot is in its turn lodged. For an appreciable time both feet remain on the ground nearly at the same spot. This position is well seen in the raccoon (series 744, Figs. 3 and 4) and in the baboon, as well as in the sloth (series 750). It affords a good example of central support. (See p. 38.)

But in fast rates the fore foot has left the ground before the hind foot reaches it, so at no moment are the two feet on the ground at the same time. Flat-footed forms are capable of retaining the two feet on the ground at the same time without the sacrifice of speed noted in the small-footed types, such as the soliped and the ungulate are compelled to submit to.

In the raccoon the fore limb may be, in some positions, used as a balance to prevent the animal being toppled forward.

Division of Parts of Limbs based upon their Movements.

The limb is easily divided into two great parts,—that from the hip to the knee and that from the knee to the toes. The movements for the most part are the to-and-fro motion of that portion of the hind limb placed to the distal side of the knee, and to that part of the fore limb beyond the elbow. The muscles which move the toes and the body of the foot arise, or tend to arise, from the lower end of the femur, while those which move the leg arise from the pelvis in great part.

In ungulates (series 682 A, Fig. 5) the entire foot may reach the ground when the animal is in rapid motion. So that the plantigrade expression is possible in an animal as far removed from that type as is the deer from the raccoon. The plantigrade foot is more flexible than the digitigrade; the distinction between the foot of the deer and that of the raccoon is great, notwithstanding the fact that the two feet may under some circumstances do the same work.

Backward Strain.

At the end of backward strain of the hind limb the fibula will act as a check to eversion. It is probably one of the uses of the fibula to thus check the eversion and enable the same bone which affords surfaces of origin to the flexors and everters to slow down the action excited by these very factors. In mammals without distinct participation of the fibula in the composition of the ankle eversion of the foot is less pronounced than in those in which it is present. Thus, in the raccoon, the fibula enters into the joint, and eversion is evident. In those instances in which the eversion is well developed yet the fibula is rudimentary, the inversion of the knee is to a corresponding degree emphasized.

In the backward strain, as seen in the raccoon, the fifth toe leaves the ground before the fourth. The leg is always rigid, but the extent to which the limb lies back of the rump varies. It is less decided in the rack than in other movements, and is not so well developed in some animals as in others. This can be seen in contrasting the movement of the horse and that of the guanaco. For the animal last named, see series 743.

When an animal retains the hind limb in backward strain and the fore limb in forward strain (the limbs being those of the same side of the body), the trunk on the same side is stretched out to the utmost and the genu-abdominal fold is made tense. (Series 680, Fig. 8.) When the backward strain of the hind limb is associated with the backward strain of the fore limb, the trunk is flexed (if such a term be permitted) and the folds just named are relaxed, and the creases of the skin on the sides of the trunk are disposed in a number of vertical lines. These creases are especially well seen in the skin of the hog. (Series 675. See p. 93.)

It is a noteworthy fact that the number and the position of the folds on the side of the hog are the same as those of the bands seen on the side of the trunk of the nine-banded armadillo, and it becomes an interesting matter for consideration whether or not the bands may not have originated in the same manner, and for some reason remained permanently.

The position of backward strain, as already announced, is most variable for the fore limb. The strain is not, however, the same for all animals. The backward inclination of the humerus is greatest in the deer. It is pronounced in the other ungulates, excepting the giraffe. In the elephant it is scarcely discernible. (See p. 91.)

It is evident that the limb that leaves the ground the last is the one which does the most work. This function probably varies at different times in the same gait, although there is nothing in the photographs which proves that such variation occurs.

The amount of forward strain is also subject to variation. The deer possesses extraordinary power to throw the fore limb forward. The degree to which the facility to accomplish this act is carried is seen in series 690, Fig. 10. It is also marked in the raccoon (series 744, 745). The movement is much less marked in the horse and the ungulates generally. The deer exhibits a much larger anterior surface at the trochlear curve of the distal end of the humerus than does any other mammal which was examined. In the deer the proportion is one-sixth of the shaft of the humerus, in the ox and horse about one-eighth.

The sudden release of the backward strain, with eversion of the foot, must naturally tend to inversion of the knee; hence a simple torsion of the entire limb is effected. (See Torsion, p. 57.)

The Manner of a Limb going "off" in the "Spring."

The nature of the "spring" is not made clear by all the photographs. It is well seen in the horse (series 642), but in the dog the immense impetus which is gained by the "spring" would appear to be produced by a mechanism altogether inadequate to the end in view. The animal rises from the ground without the source of the necessary power to accomplish this act being evident. In the rabbit and the kangaroo it is easily understood how by the simple extension of the tarsus from an acutely flexed position, followed by the flexion of the digits, the body is sent upward and forward; but the leg of the dog is already extended at the time that the "spring" is initiated. If, as is asserted by practical men, the horse is prepared for the "spring" by the same succession of foot-falls as in the gallop or run, it is only necessary to look upon the last limb which leaves the ground as though it were engaged in preparing the leg for a recover from an average stroke. Since the eversion of the foot presses the inner border against the ground, the inner aspect of the limb at the knee is directed inward and subjected to strain. But the force generated by this strain is unexpended at the time that the foot is raised from the ground, so that the limb might be compared to a coiled spring which has not lost its power, and the stored-up energy is engaged to propel the body forward the moment that the inner border of the foot leaves the ground. While acknowledging that this is an hypothesis, it remains the best of the surmises which occur to the observer.

It has already been seen that when the limb is in the position of arrest and the momentum carries the body beyond the perpendicular line it is thrown into "backward strain." The instant that the strain begins the knee is seen to move outward and the hock to move inward. The parts of the foot below the heel remain unchanged. The impact of the structures of the limb is thus impaired in backward strain. It is well known that in the pentadactyle forms the foot can be readily rotated at the mediotarsal joint, and it is a reasonable assumption that it is at this joint that the distal part of the limb moves when the entire limb rotates outward. The femur, the bones of the leg, and the astragalus act as one factor, and the calcaneum and the remaining bones of the foot as the other factor. The socket for the proximal motion occurs at the hip, and that for the distal at the concavity of the scaphoid bone. There is also considerable motion between the calcaneum and the cuboid bone and between the calcaneum and the lower end of the fibula, if this bone is present, or with the outer end of the tibia if it is absent. Outward rotation of the main portion of the limb carries the calcaneum slightly inward by reason of the articulation between the calcaneum and the bones of the leg. Facets are here present in most terrestrial mammals. In the wombat the articulation is evident. It is present in a rudimentary form in man.

The outer surface of the calcaneum of the bear is marked by a stout roughened ridge as it enters into articulation with the fibula. In the dog the surface is a small embossment, which probably is in contact with the fibula only at the time of the backward strain. In a single old dog examined the same ridge is present as in the bear. A similar ridge which developed under the stimulus of diseased action is seen in the skeleton of the tiger in the Museum of the Academy of Natural Sciences of Philadelphia.

As the knee is rotated outward the outer border of the foot is slightly inverted. This disposition is opposed by the peroneous longus muscle which everts the foot. Coincident with the inversion the external crucial ligament becomes tense and the tendency to torsion is checked.

The degree of outward rotation may be found to relate to the swing of the trunk. In the elephant and the camel the rotation is decided. In the goat it is apparently absent. It is less marked in the horse than in the cat.

The Act of Turning Round.

In the dog the act of turning is accomplished in the following manner:

The hind feet remaining fixed, the body moves in front. It is thrown laterally on one fore limb, so that this member becomes more and more oblique to the line of gravity. At a time when a line drawn upward through the supporting foot will lie entirely free from the trunk the opposed limb has been placed in the line of gravity and a new support assured. In the raccoon (series 745, Fig. 17) the outer border of the foot is used for turning.

Other figures which exhibit the phenomenon of the act of turning are those of the dog, the horse, the tiger, and the lion.

On the Effect of Positions of the Limbs on those of the Head, the Neck, and the Trunk.

When the forward movement of the fore limb is the most conspicuous feature of the figure the head is erect. When the position of backward strain is the most conspicuous the head is depressed. When the head is being erected the neck becomes convex in front and concave behind. When the head is depressed the neck is concave in front and convex behind.

The movements of the head and mane can be satisfactorily studied in all rapid gaits of the horse, especially in the trot and the gallop. The mane flies backward as the head descends and clings to the neck as it ascends.

The series of figures (649 A) of the trick-horse "Hornet," rocking to and fro, is of especial value in studying correlations of head position to limb movement. (See p. 93.)

The Influence of the Body on the Shape of the Foot.

In the goat (series 677) the outer parts are larger in Figs. 10, 11, 12, and the inner are larger in Figs. 1, 2, 3, 4, and 5.

In the goat the chief weight comes on the inner toe. In the doe (series 641) the two toes separate at equal angles to the axis of the leg. In the ox and the camel no difference is perceptible. Profile views of the lioness and of the elephant are interesting in this connection. In the capybara the inner border of the foot bears the weight.

Value of the unequal Lengths of the Toes.

In a plantigrade animal like the raccoon the transition from the plantigrade to the digitigrade form will bring the short toes to the ground in the digitigrade position at times when the longer toes are yet on the ground their entire length. This position is well seen in series 744, profile, Fig. 19. As the foot passes still farther towards the tips of the larger toes the short toes leave the ground. At first they are in the axis of the metapodium, but afterwards they are thrown backward, and near the end of backward strain appear to be abducted. (See series of the raccoon, 744, Figs. 9, 12, and 13. See p. 73.)

On Obliquity of the Outer Toes.

The oblique movement will be essentially the same in all instances no matter what the number of functionally active toes may be. A straight toe on the limit of the inner border and a deflected toe on the lateral aspect are always seen. In man the first toe is straight and all the others are inclined outward. In the goat, which has but two functionally active toes, the outer of the two is deflected.

The action of the toes must be studied in connection with the outward rotation. It is evident that the outward rotation must be at an end, or near an end, before the digits are taken from the ground. (See rotation in elephant, p. 90.)

THE ACTION OF THE LIMB WHEN GOING OFF THE GROUND AND WHEN IN THE AIR.

The first movement noticed in the limb after it is beyond the centre of gravity is the flexion of the foot. In the horse the hoof is thrown backward, and the under surface of the foot is directed backward, the heel being raised first. The sole is next directed upward. In animals possessing more than one functionally active toe the toes are quickly adducted in the air, so as to offer the least resistance to the impetus of the entire body. Associated with the above, a pronounced flexion of all parts of the limb occurs excepting at the hip, where the movement is slight. A movement of the thigh towards the trunk is faintly discernible. In animals possessing long thigh-bones, such as the elephant, the movement is more decided than in hoofed animals. The same remarks are applicable to the movements of the humerus. The degree to which flexion is carried is more marked in the young than in the adult, and in terrestrial than in arboreal creatures. In the sloth (*Cholœpus*) flexion is absent, the limbs being advanced by a swinging motion at the shoulder and the hip. (Fig. 11, p. 83.) In rapid gaits the unaided eye receives the impression of backward movement, but fails to be impressed with forward movement. It may hence be inferred that the former is a quicker movement than the latter.

In the fore limb the last state of extension of the forearm answers to the action of the extensors of the carpus and of the digits. In the less delicate movements of the hind limb the muscles which extend the tarsus and the toes move the foot with less precision, and, it is likely, with less speed. The first stage of the recover is a deliberate act. Beginning at the toes, it ends at the hip. The second stage is quicker than the preceding and is more abrupt. Beginning at the hip, it ends at the toes.

Proposition that the Foot strikes the Ground by the Outer Border.

This movement most probably takes place in a constant manner in all quadrupeds. Huxley* describes the act in the chimpanzee as though it were peculiar to that animal. In the horse the movement is well seen in series 601, Figs. 3 and 4, left hind foot. It is also seen in Fig. 4 of the right foot of the same series. The action is well seen in series 581, Figs. 10 and 11. (See pp. 42, 84, 88.)

The foot in all animals excepting the horse (and even in this single-toed form the movement of the foot is in nearly all essentials the same) is carried forward in semipronation. After the foot strikes the ground on the outer border pronation begins, and is completed by the time the perpendicular line is reached. The foot leaves the ground by the inner border (the toes being successively abducted), so that the pressure of the body is borne from without inward across the foot. (See Fig. 4, p. 73.) The foot is always everted as it leaves the ground. In a plantigrade animal, as the raccoon, the foot is carried during the last part of recover nearly parallel to the plane of support. In the rapid motion of ungulates and of the horse the foot may actually touch the ground nearly to the hock. (See p. 43.) In backward strain the hock or heel is gradually raised, and at the end of strain the animal is seen touching the ground by the tip of the inner functionally active toe. In the horse the foot leaves by the tip of the hoof. It is likely that the degree of impact of the outer border of the foot will be found to correlate with the degree of development of the calcaneo-sural joint,† since the weight must be carried along the outer border to the rest of the limb. At the end of backward strain the limb from the knee distally is in the same line.

The entire series of changes from semipronation to pronation, as studied in connection with the transfer of weight across the foot, is well seen in the tiger (series 729 and 730; see also p. 85).

The Eversion of the Foot.

The moment flexion begins eversion is established, and the limb becomes angulated outward at the ankle. The main axis of the proximal facet of the astragalus is correlative with the degree of

---

* *Medical Times and Gazette*, April, 1864, 398.

† A name proposed for the joint existing between the fibular process of the calcaneum and the fibula, or the tibia.

this obliquity. The angulation is most pronounced in the horse, less so in the ox, and scarcely, if at all, in the hog. (See p. 55.)

It is evident that the more the toes are separated the better the limb is supported. It would appear to be one of the uses of the digits that they are capable of moving so as to increase the width of the under surface of the foot. Such an increase of width ensues from two causes,—first, from the pressure outward of the foot against the ground; and, second, from the traction of the digits by the action of muscles. The muscular action is most pronounced. It is well defined along the outer border of the foot. As the pressure shifts across the foot from the outer to the inner border, the digits leave the ground and are lifted and abducted as they are relieved from duty. In the capybara the movement of abduction begins before the limb reaches the ground (series 746, Fig. 6). Just as the foot is lifted the inner border is turned outward and the sole is disposed to the air in the position of eversion,—that is to say, the sole, while directed outward, is so arranged that the inner border of the foot is directed downward and the outer border upward. As the foot is everted the toes become abducted.

Eversion is well seen in the horse (series 622 and 633). It is present, but less marked, in the deer (series 681). A good study can be seen in series 619. In the slow movement of the walk the feet are not everted at the moment of the removal of the weight, but after the feet are well in the air. This is especially noted in series 581.

In series 594 the figures of foreshortening show that when, as in Figs. 3 and 6, the fore foot is lifted, the foot moves towards the median line, and when the hind limb moves it swings off from the trunk. It is likely that the greater width of the pelvis over that of the withers compels the hind limb to describe an excursus outward.

That the amount of motion inward under the body is susceptible of modification is evident from the fact that among fast pacing-horses the feet are brought inward, in order to reach the same spot on the ground. Since the fore limb is the more mobile of the two, it is likely that in the process of adaptation* the in-

* The word "adaptation" is here used in the sense that a terrestrial animal possesses the power of modifying to a slight degree the movements of the limbs according to the character of the ground it is moving upon. (See p. 96.)

ward movement is secured by the front limbs, and that the hind limbs remain the same. (See p. 42.)

In the sloth the foot is lifted as well as swung. (See Fig. 11, p. 83.) It would be interesting to ascertain if the ape also lifts the limbs in swinging.

All elements of the swing of the hind limb appear to be lost in the "jump" if the two hind feet act synchronously.

In the camel the fore foot in backward strain answers to a vertical line which intersects a point as far back as the middle of the posterior dorsal hump, while the hind limb advances forward as far as the concavity between the humps. (See p. 38.)

The Degrees made by the Limbs at the Extremes of Forward and Backward Movement.

The fore limb when at the extreme of forward movement presents an axis which is very oblique as compared to the limb when in the perpendicular position. Again, when in the extreme of backward movement it is again oblique to the same perpendicular position. It has been thought of sufficient interest to record the exact degrees presented by these extremes of movement in the fore limb and to contrast them with similar movements of the hind limb.

By the use of the word plus, or the plus sign (+), a deviation to the right of the vertical is indicated. In like manner minus, or a minus sign (—), indicates a deviation to the left.

In the elephant (series 733, Fig. 13) the forward strain of the fore limb is 25° minus. The backward strain is 35° plus. In the hind limb (Fig. 10) the forward strain is 15° plus and the hind strain (Fig. 9) is 45° minus. In the lioness (series 728, Fig. 3) the forward strain is 50° and the backward strain 20°. In the hind leg the forward strain is 50° and the backward strain 50°.

In the dog (series 707, Fig. 1) the fore foot at the end of backward strain, but as yet on the ground, is in an axis of 10° minus; the rest of the limb being 20° plus nearly. In Fig. 2, the foot having left the ground, marks 80° minus, nearly, while the rest of the limb remains nearly the same as when on the ground,—namely, 18° plus. In the hind leg, as in Fig. 5, the foot when at the end of backward strain gives an axis of 90° minus; while that of the metatarsus gives 10° minus; and that of the thigh is less than 5° from the vertical. In Fig. 6, the foot has left the ground, and with the metatarsus and leg to the knee, become deflected to 40° minus, while the thigh remains the same,—namely, less than 5° plus. In series 690 (*Cervus dama*, fallow deer) the anterior oblique is 140°, and the posterior 10°.

The method here accepted of determining the degrees of the deflection of the different parts of the limb is sufficiently accurate for all purposes except in the case of the thigh, where the slight movement of the entire mass forward is an imperfect index to the deflections of the femur.

It will be seen that the fore limb moves as a whole, while the hind limb varies in each of its segments. It is true the fore limb presents an angulation between the axes of the digits in hoofed animals and the rest of the limb, but on the whole the limb may be compared to a prop. (See pp. 41, 62.)

In a general way it may be said that the forward strain of the fore limb is parallel in the lioness to the forward strain of the hind limb; but the deflection of the hind limb in forward strain is two and one-half times greater, as compared with that of the fore. In the elephant the forward strain of the fore limb is greater than the forward strain of the hind limb by ten degrees; and the backward strain of the fore limb is ten degrees less than that of the hind. The general trend, therefore, of the two limbs in animals so remote as the tiger and the elephant presents contrasts which in their way are as great as the anatomical peculiarities of the animals themselves.

The correlations of these deflections with the trunk are doubtless exact, but no systematic effort has been made to define them. A vertical line drawn upward through the foot and made to intersect the vertebral column might be made available in the formation of an index of importance. It is noticed that in the photograph of the lioness this vertical line corresponds to the second of the minute dorsal elevations. As a general rule, it may be said that the line intersects the vertebral column at a point answering to a distance of twice the breadth of the withers. Again, in the lioness the forward strain of the hind foot yields a point by which the vertical line can be drawn upward which intersects the vertebral column at the last of the dorsal marks.

While the purpose of the forward strain in the claw-footed

animals is to secure a purchase on the ground by which the trunk can be pulled forward, in the hoofed animals it exhibits another, —namely, to protect the massive head. (See succeeding section.)

The following measurements have been made of the hind limb of the horse at end of backward strain in the rack, the pace, the trot, and the gallop:

Rack.	Pace.	Trot.	Trot.	Gallop.
		Time, 17.	Time, 19.	
590, Fig. 2	591, Fig. 9	607, Fig. 16	609, Fig. 2	631, Fig. 3
Foot, 2°+	Foot, 0°	Foot, 0°	Foot, 1°+	Foot, 0°
Metat., 8°+	Metat., 21°+	Metat., 21°+	Metat., 30°+	Metat., 40°+
Leg, 40°+	Leg, 50°+	Leg, 40°+	Leg, 40°+	Leg, 60°+

Extent of Forward Movement of the Anterior Extremity.

The position of the hoof of the fore foot in forward strain varies with respect to the head in different ungulates. In the heavily-built forms, as the ox, it reaches the occiput only, in the Indian buffalo (series 701) the ear, while in the deer (series 688) it appears to advance even beyond the muzzle. (See p. 38.)

This arrangement doubtless correlates with the massiveness of the head, when the fore limb comes to the rest beneath the head, as in the Indian buffalo and allied animals, and with speed, when it reaches the ground in advance of the centre of the head.

The extreme position of the hind limb after it is taken off varies greatly, as indeed it also does in the limb just prior to its being lifted from the ground.

The changes in the relative positions of the fore foot when off in the various gaits are important. In the dog and the doe the entire extremity is acutely flexed. In the adult deer it is extended, as is also the case in the cat. (See p. 92.)

The hind foot appears to retain its position for a longer time than the fore foot,—that is to say, in the fore limb the impression is given that it is preparing to leave the ground soon after it passes its vertical, while in the hind limb the propulsive power is expended through the entire period of the "stroke," and is as pronounced in the positions back of the vertical as in any other.

The toes of the fore leg are flexed quickly, even in Echidna and Solenodon. So far as observed, all mammals flex the toes of the fore leg not only quicker than the hind, but more completely.

The adduction of the toes after the removal of the weight of the body does not appear to be the result of a mechanical necessity, for it is subject to variation. It is less prompt in the fore limb than in the hind. At times when the hind foot is raised it tends to be abducted. An example of such retention is seen in series 594, Fig. 5.

"Slowing Up."

The descent of the limb must have an appreciable effect on the momentum of the body by presenting a surface of resistance to the air. This occurs in the manner made familiar by the bird in the act of depressing the tail-feathers to arrest flight. Great breadth of limb is therefore an unfavorable condition for rapid locomotion. The figures of a cat jumping, as seen in series 719, Figs. 1, 14, and 15, serve to illustrate the action.

Position of Feet in the Last Stage of Recover.

The feet in forward motion are carried in the position of semi-pronation during the first stage of the movement. This act is the best adapted to rapid action, since it presents the smallest surface of the limb to the resisting air. (See p. 49.)

The Angulation of the Limb at the Ankle due to Direction of the Trochlear Axis of the Astragalus. (See p. 50.)

Among the more pronounced correlations which exist between the contour of the limb and the deeper structures is the shape of the ankle. This is seen in the horse, the ox, and the hog. If the astragalus of each of these animals is examined, it will be found that the deflection of a line drawn through the upper or pulley surface forms an angle with the axis of the limb of varying degrees in the horse and in the ox, but none in the hog. The production of such a line in the limb whose leg- and foot-bones are in the positions assumed in life will show that the produced line of the surface in the horse intersects the metatarsus twenty centimetres below the proximal end of the bone, in the ox at eleven centimetres, and in the hog the axis is parallel with the tarsus.

With these facts before us and the photographs examined, it will be seen that in the horse (series 576), the figure of the animal being seen foreshortened from behind at ninety degrees, the right hind foot is barely off the ground, and has not lost the position of the ankle when seen in backward strain. The produced axis of the leg will intersect the axis of the metatarsus at the position which corresponds to the production of the axis of the astragalus-pulley in the skeleton.

In the ox (series 671, Fig. 12) the rear view of the left hind leg exhibits a totally different inclination. The foot is on the ground, but is about to leave it, and the produced line is much nearer the vertical. In the hog, as shown in series 674, Fig. 4, the axis of the astragalus as far as can be ascertained is also that of the leg.

Training and Age as Factors of Disturbance.

The difficulty in studying the actions of the domesticated animals, especially of the horse, is owing in the main to the fact that the training to which the animal has been subjected modifies his movement. The draught-horse comes down on the tip of the foot instead of the heel, as is the case with other quadrupeds. In like manner the movements of the well-bred horse are influenced by the skill of the driver. In the gallop (series 624) it is evident that the rider is at fault in the management of the horse and disturbs his action.

The trot and the fast gaits, such as the run and the act of the leap as taken by the horse trained for the circus, are unnatural, and belong to the class of the acts of the acrobat and the contortionist as seen in the human movements. (See p. 66.)

The fact that the horse in racing can place each of the feet in succession in the same spot is in this relation unessential, and need not militate against any conclusion drawn from slower and more natural movements. The same remark is applicable to the fact that the fast horse, even in the walk, will place the hind foot on the ground in advance of the fore foot.

The age of an animal will modify the movements. An old horse, as already seen, will place the tip of the hoof to the ground instead of the heel. In the same class the hind foot is advanced to a less degree than the young. In the colt, on the other hand, the forward strain is the greater. In the fawn, as already remarked, the feet are more flexed than in the adult deer. An untrained adult feral animal is, all things remaining the same, the best form which should be taken for study.

The weight of the rider being thrown forward on the fore legs may modify the gait. That the depression of the head in the forward movement of the legs is not dependent upon this circumstance is shown in the figures of the rocking horse (series 649 A). An additional reason is here presented why, when practicable, a form taken for study should be feral. (See also remarks on the trot, p. 66.)

Torsion of the Trunk.

An animal in thrusting out a limb from the body may be compared to a boxer making a thrust with his arm. With each lunge the body is turned in the direction in which the lunge is made, and the impetus of the body-movement is added to the force of the blow. The twist that the body describes is checked by the planting down of one of the disengaged feet.

In the dog (series 707), while both hind limbs are off the ground and the body is being vaulted on the single fore limb, the entire posterior part of the body is deflected from the line of the main axis of the vertebral column. This motion is certainly dependent on the bending of the column, and is probably an example of torsion. If unchecked the motion would "twirl" the body round on itself.

But since the impression of the foot on the ground remains sharply outlined, it is probable that the "twirl" is greater in the proximal than in the distal joints, and is entirely lost by the time that the parts of the limb are reached which rest upon the ground. Thus the first impetus towards deflection, while originating in the vertebral column, is gradually transferred to that limb which at the time is serving as a prop, and the force of the movement is dissipated as the animal is carried forward.

In animals possessing a rigid vertebral column the torsion which corresponds to the movement above described is not marked. But the disposition of the body to incline towards the side which perfects the forward strain—*i.e.*, delivers the blow—is the same. Such inclination if unchecked would throw the body off the straight line. The check is secured by the hind foot of the opposed side of the trunk coming on the ground.

Series 707, Fig. 6, shows to advantage the disposition exhibited by the dog to twist the trunk during progression. This tendency is never seen in the finer varieties of the animal.

The Act of Kicking.

In series 658, which exhibits the kick of the horse, Fig. 1 shows the hind foot as not quite extended. When the kick is complete the foot is fully extended, as seen in Fig. 2 of the same series.

The Position of the Fore Limb as to the Axis of the Trunk.

An inclination exists for the foot to rest on the ground directly in the middle line of the body; but the foot is not brought to the ground in this position, but to the opposed side,—that is to say, it crosses the median line, and comes down on the right side of the line if the left foot is considered, and to the left if the right side is considered. Such disposition is seen in the photographs of the dog (series 703, Fig. 2, and series 704, Figs. 5 and 6). The motion described is best seen in the heavy type of animal. It disappears in the slender breeds, as the greyhound. The same remark is applicable to the horse, the crossing of the fore legs being seen only in the heavy breeds.

The foot which comes to the ground at the median line will gradually be drawn to the lateral portion of the trunk, as seen in series 709, Figs. 4 and 5. In this position it leaves the ground. The prop movement must from this circumstance be less efficient towards the end of the vault than it was at the beginning.

The crossing of the fore legs is admirably well seen in same series.

Depression of the Heel.

Complete forward movement of the limb tends to depress the heel, or, *per contra,* the forward movement of each extremity tends to draw the toes away from the plane on which they are to fall (series 601, trotting-horse).

Inward Rotation.

After the hind leg leaves the ground the limb is disposed to rotate inward, an act well seen in the horse and the elephant

(Fig. 1). The ilio-psoas would have an apparent effect in preventing this. It may be surmised that this muscle is a check to the excessive inward rotation, or the muscle may not contract until later, and prove to be the chief factor in the second stage of the recover. (See p. 89.)

FIG. 1.

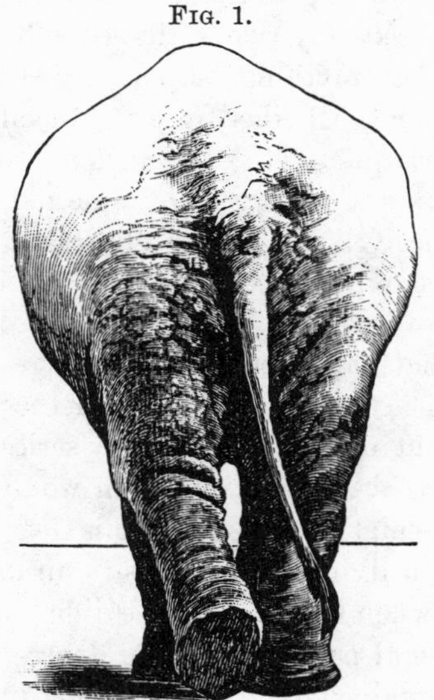

The figure is designed to illustrate the disposition for the limb to rotate inward. The inner border of the uplifted foot is oblique, and the leg is inclined towards the median line of the body.

THE GAIT, OR THE SUCCESSIONS OF FOOT-FALLS.

The order in which the feet comes to the ground would be a simple matter for study if the animal moved constantly at a given speed and gait. It is rational to assume that the movements are in part automatic, and the alternate motions of the right and left limbs, or the motions of both the left limbs alternately with both the right limbs, would insure a succession which could be premised.

Motions are as a rule rhythmical. One expects the serpent to undulate the trunk after a regular method, no matter what the speed of the animal may be. In flying and swimming, so far as is known, the movements are synchronous and constant.

In the terrestrial movements of the quadruped, however, such is far from being the case. This is owing to a variety of causes.

So far as the photographs in the Muybridge series are concerned the order of succession of foot-falls could have been improved—that is to say, could have been more accurate—if the movements in the record of each gait had been of exactly the same character. But in point of fact, owing to the practical difficulties surrounding the subject, scarcely any two of the series begin and end in the same manner, nor has anything been recorded of the mental conditions of the animal, whether it was tractable or intractable, whether excited or quiet, etc.,—conditions so essential to the manner in which an animal may determine its motions.

If it can be conceived that a perfectly tractable and composed animal had been moving in a circle, and the cameræ instead of being arranged in a line had been grouped in a central cluster in such a manner that they could secure correct pictures of the moving form in the same way as in the method actually adopted, it is likely that out of such an endless series a uniform set of pictures might have been secured which would have given completeness of representative actions. It is reasonable to suppose, that the motion of a living creature, as in an unvital mechanism, when starting and when halting, may be different, and may present contrasts of the several parts to a greater degree than when studied at a regularly maintained rate, whether this be a high or a low one.

Each position of the foot when three feet are down embraces the laterals with one of the opposed feet in addition. Thus, in series 738, Fig. 1, in addition to the right laterals the left hind leg is on the ground. But in none of the pictures were the left laterals detected in which the right hind limb was "on." It is probable that the diagonal heterochiral grouping is a weaker form of support than the lateral, and is used as an expedient to shift the laterals from the right to left and back.

In the use of the laterals, in all gaits the feet as they approach the end of the time at which they are "on" are at the side of the trunk, while the feet in use as diagonals always remain beneath the trunk.

The facts that no variety of the deer ever paces, that the mule-deer is the only variety that bounds,—*i.e.*, that all the feet leave and strike the ground together,—that the Canadian deer soon becomes fatigued in the run, while this gait is the one longest assumed by the Virginia deer,* point to the conclusion that natural gaits are correlative with structural peculiarities; and while quadrupedal movements are based on the same plan, they are subject to modifications in animals otherwise closely related.

Gait may be modified, if not controlled, by the proportion of the length of the limbs to the trunk. The hippopotamus walks in perfect synchrony, so far as the unaided eye can detect. The fat pig can walk, but cannot gallop. The lean pig, on the other hand, can gallop.

The fact that the giraffe is the only animal which is a pacer for both slow and fast movements cannot be easily disassociated from the great length of the limbs.

The subject of the gait can be studied by *observation* of the animal as a whole, by *numerical analysis* of each foot as it is either "on" or "off" the ground, by numerical analysis of each combination of any number of feet, and by *graphic analysis*, by which means tracings of the feet in combination are secured.

For convenience the subject is divided into the following heads:
The Synchiral Gaits.
The Heterochiral Gaits.

Synchiral Gaits.

The Synchiral gaits embrace the canter, the gallop, and the run.

In the canter the momentum of the body is not sufficient to enable both the fore limbs to act as props until the hind limbs again are in the position to urge the body forward. Hence the animal is observed to come down upon a fore limb after being in the air.

In the gallop the fore limbs act successively as props before the hind limbs come down. The run is the variety in which the limbs attain the greatest possible amount of strain, and in which the quick succession of the paired feet is absolute.

The "jump" is an incident in the run. In jumping a hurdle the horse, in series 641, appears to come down on two fore feet at the same moment. In series 637 the horse comes to the ground on the left fore foot, and is instantly followed by the hind foot of the same side.

---

* J. D. Caton, The Antelope and Deer of America, p. 270.

The Gallop. (See pp. 41, 53.)

In series 632, the front legs are "on" and "off" equally, each being "on" 2 and "off" 10 in a series of 12. While the hind legs are unequal, the left leg being "on" 5 and "off" 7, the right is "on" 4 and "off" 8.

In series 631, the fore feet are equal, being 3 "on" and 9 "off;" the hind legs are again unequal, the left being "on" 3 and the right 4.

In series 633, the feet are all moving equally. Each foot is "on" 3 and "off" 9.

It will be seen that the fore feet move evenly for the front, and in two of the three series twice unevenly and once evenly.

When the horse is going at a moderate rate of speed the greatest distance between any two of the impressions made on yielding soil answers to the interval between the anterior of the impressions of the hind feet and the first of the fore feet. When the animal is going at a high rate of speed the greatest distance is between the fore foot by which he leaves the ground on the leap and the hind foot by which he alights.

In the canter the impressions made by the hind feet on the soil are in the same relative positions as in the gallop,—*i.e.*, they follow one another at wide intervals. The fore feet, however, yield impressions which lie close alongside one another,—at about the same place.

In contrasting the two gaits, it is evident that in the canter the hind feet act with such ineffective energy that the fore feet are compelled to follow each other in rapid succession to prevent the fore part of the trunk from falling. In the gallop the momentum of the body carries it easily over both fore legs, which can leisurely in succession act the part of vaulting-poles.

The gallop as well as other gaits show irregularity.

In series 680, oryx, this is reduced to a minimum. In this interesting series all the limbs excepting the right fore limb are "on" and "off" equally. Even in the right fore limb the departure from uniformity is not great, the order being "on" 4 and "off" 6.

In the fallow deer (series 682) the limb-movements are uniformly "on" 5 and "off" 11, except the left hind foot, which stands "on" 6 and "off" 19.

In the Virginian deer (series 683) the right fore limb is "on" and "off" 13. The left fore limb is "on" 9 and "off" 15. The right hind limb is "on" 8 and "off" 16, while the left hind limb is "on" 10 and "off" 14.

Intervals between the Fore and Hind Synchiries.

The interval is larger in the horse than in the ungulates. It is exceedingly small in the guanaco, series 743.

The horse may assume the lateral position in the gallop, as is seen in series 632, Fig. 11. In this figure, as well as one in series 612, the animal uses the lateral position immediately after reaching the ground. That the assumption does not depend on speed is shown in the canter. In series 616 there is lateral support in twenty-four pictures, in series 617 two in twelve pictures, and in series 619 one in twelve pictures.

HETEROCHIRAL GAITS.

The heterochiral gaits embrace the walk, the trot, the rack, and the pace.

The Walk.

This gait has been especially studied from the figures of the series of the horse, the buffalo, the ox, the cat, and the raccoon.

In the study of the walk by the numerical method it is found that in the horse (series 579), while the hind feet are synchronous, the front feet are asynchronous.

In the *horse* (series 574) the hind limbs are practically as asynchronous as the fore, while synchrony exists in the diagonal between the left hind limb and the right fore limbs.

In the same series of twenty-four pictures the following is the order of the feet that are "on" and "off:"

Rh "off" 14, "on" 10.
Rf "off" 9, "on" 15.
Lh "on" 13, "off" 11.
Lf "on" 14, "off" 10.

It will be seen that the right hind and the left fore legs are synchronous.

Series 579. Rh "on" 8, "off" 4.
Rf "on" 7, "off" 5.
Lh "on" 8, "off" 4.
Lf "on" 9, "off" 2.

In the walk of the *buffalo* (series 699) the order is as follows: Rh "on" 18, "off" 6; Rf "on" 16, "off" 8; Lh "on" 10, "off" 14; Lf "on" 15, "off" 9.

Nothing here is synchronous; the greatest discrepancy is seen in the diagonal and the least in the lateral feet.

In the *ox* (series 670) the difference between the two fore legs alone is great: Rf "on" 11, "off" 1; Lf "on" 8, "off" 4.

In the *cat* (series 720): Rf "off" 20, "on" 4; Lf "off" 15, "on" 9; "Rh "on" 9, "off" 15; Lh "on" 9, "off" 15.

Here all feet are acting synchronously excepting Rf. Between the fore limbs there is the most marked difference. The left fore limb remains on the ground over twice as long as its fellow.

In the horse (series 576) the diagonal Rf and Lh are "on" 8, "off" 4; while the opposed diagonal is Rh "on" 11, "off" 11; and Lf "on" 6, and "off" 6.

The walk in the *raccoon* (series 744) shows that the fore leg moves slower than the hind leg. When the momentum of the animal is increased the fore leg attains the same speed of rapid change as the hind.

The rate of the "on" and "off" of the fore feet is ½ to ½, while in the horse (series 579) the rate is ⅔ "on" to ⅓ "off."

The left fore foot is "on" the ground in 1, 2, 3, 10, 11, 12, 13, 14, 20, 21, 22, 23, = 12.

The left fore foot is "off" the ground in 4, 5, 6, 7, 8, 9, 15, 16, 17, 18, 19, 24, = 12.

The left hind limb is "on" in 1, 8, 9, 10, 11, 17, 18, 19, = 8.

It is "off" in 2, 3, 4, 5, 6, 7, 12, 13, 14, 15, 16, 20, 21, 22, 23, 24, = 16.

Thus the left fore foot is in the act of supporting and propelling the body for the same length of time as it is engaged in preparing for a second stoke. In the left hind foot the time of the foot when engaged in support or propulsion is but one-half that of the time when it is engaged in the air.

Combinations in the Walk.

Out of the number of combinations which it is possible to secure from the feet when on the ground the following is the arrangement most frequently seen in the act of walking: Two hind feet alone; two fore feet alone; two hind feet and two fore feet. The combination one hind foot and two fore feet has not been seen, though no reason can be urged for the non-occurrence. (See especially series 744.)

Support of the Vertebral Column in the Walk.

The lateral disposition of the feet in the walk supports the vertebral column at the ends, and the figure of the trunk and limbs compares with a truncated triangle whose base is on the ground. In the diagonal disposition the opposed feet meet at a point near each other under the centre of the trunk (when the other feet are "off"), and the figure compares with that of an inverted triangle whose curved base is directed upward.

The animals which assume the diagonal central support are short-legged creatures. The best examples are the raccoon and the baboon. (See p. 72 and p. 75.) The approximation to it is seen in the capybara. It is never seen in the ungulate.

The Trot.

In the trot the succession of foot-falls is inclined to synchrony from the circumstance that the motion becomes more constant as the animal increases in speed.

Thus, in the trot in a series of twenty pictures,

Rh is "off" 15.
"on" 5.
Lf is "off" 15.
"on" 5.

In 593—another series—the order is as follows in a series of twelve pictures:

Rh is "on" 5.
"off" 7.
Lh is "on" 9.
"off" 3.
Rf is "on" 4.
"off" 8.
Lf is "off" 9.
"on" 3.

The left feet have moved in synchrony. The right feet show a disposition to remain on ground longer, and, therefore, are doing more work.

In the trotting-horse the fore leg which is about to leave the ground is apt to interfere with the hind leg of the opposite side as it is advancing to come to the ground. In order to avoid this mishap the animal is trained to lift the fore legs to a degree which is not required in the walk. The motions of the animal suggest that the act of trotting as developed in the horse for fast gaits is artificial and is not especially useful for studies in locomotion. It has the same value as the study of the motions of a "trick-horse." (See p. 56.)

The Rack.

That the rack is not a uniform gait is seen from the following statements:

In 594, in a series of twelve pictures, the animal stands on one foot alone once; is in the air in six; and on laterals in five.

In 595, in a series of the same length, one foot remains "on" in three; laterals are "on" in six; and the animal is "off" in three.

The transition between laterals is well seen in the series 593. In the walk at least the transition between the laterals and the diagonals is effected by the hind limbs.

In racking the impressions made by the hoofs on the ground are in regular alternation of front and hind feet.

Pacing in the horse is a delicate motion, since any irregularity of the ground interferes. The giraffe is a natural pacer. The pace is often accepted by the dog.

Transitions and Contrasts of Gait.

The walk is the basis of the trot and the amble, which is indeed nothing but a running walk. The gallop leads to the run. The run may be called an extended gallop. The amble and the rack are trained forms of the gait. The walk often breaks directly into the amble, and this into the gallop.

The difference between the trot and the pace, as noticed in the effect the positions of the limbs make on the eye, is but slight.

In series 602, Fig. 4, the fore and hind legs on the right side are both in forward strain. The fore leg comes down first, and is instantly followed by the hind. If the two had come down together, we should have had the initial step of the pace.

The trot and the walk are closely allied in all essential particulars. The manner in which the imprints of the feet are made on the ground is in both gaits the same. The front and hind feet of heterochiral pairs impress the ground at the same places. The intervals between the sets of heterochiral impressions are in the walk about thirty-eight inches; in a small horse they may be twenty-five inches. The contrasts between the figures in the photographs of the horse at the walk and at the trot are very pronounced in the profiles. (See series 584 and 601.) But in the foreshortenings the two gaits present appearances which are very similar. The reàder may compare Figs. 4 to 8 with profit.

The trot and the walk are both heterochiral, and vary in the manner of lateral or diagonal positions of limbs.

The chief difference between the impressions made on yielding ground of the feet in the trot and the walk is seen in the relative distance between them. The impressions are much farther apart in the trot than in the walk.

Gaits are more variable in the slow than in the fast varieties. This is inevitable, since the fast gaits depend upon uniformity. Dr. Cryer, in conversation with the writer, states this as an impression.

The Laterals in the Walk and the Rack contrasted.

The lateral position as seen in the walk is essentially different from that seen in the rack. In the walk the fore foot is in forward strain, while the hind limb is in backward strain. In the rack, the fore foot being in forward strain, the hind foot is in the same position, so that the limbs are parallel to each other.

In Procyon (series 744), the diagonal heterochiry shows that the fore and the hind leg, say of the right side, are both in forward movement, while the fore and the hind leg of the left side are in backward movement.

In lateral heterochiry the disposition is reversed: when the fore limb is in forward movement the hind limb of the same side is in backward movement.

The *graphic method of analysis* embraces the following consideration: The essential lines in a given movement have been drawn and arranged in the same order as in the photographs. (See series 577.) As the walk is the most complicated of the gaits, it has been selected as an example of this method. The advantage is evident, for the observer is relieved of the action of but one pair of feet. The eye can pass easily from the combination of left hind and right fore feet—as in the line to the left—to that of the right hind and left fore feet of the lower lines to the right; and the plate itself being before the observer, the study of the succession of foot-falls becomes an easy matter.

The Graphic Method of Analysis.

The Limbs of the Horse at a Walk in Diagonal Heterochiry.
(Series 577. See pp. 64, 67.)

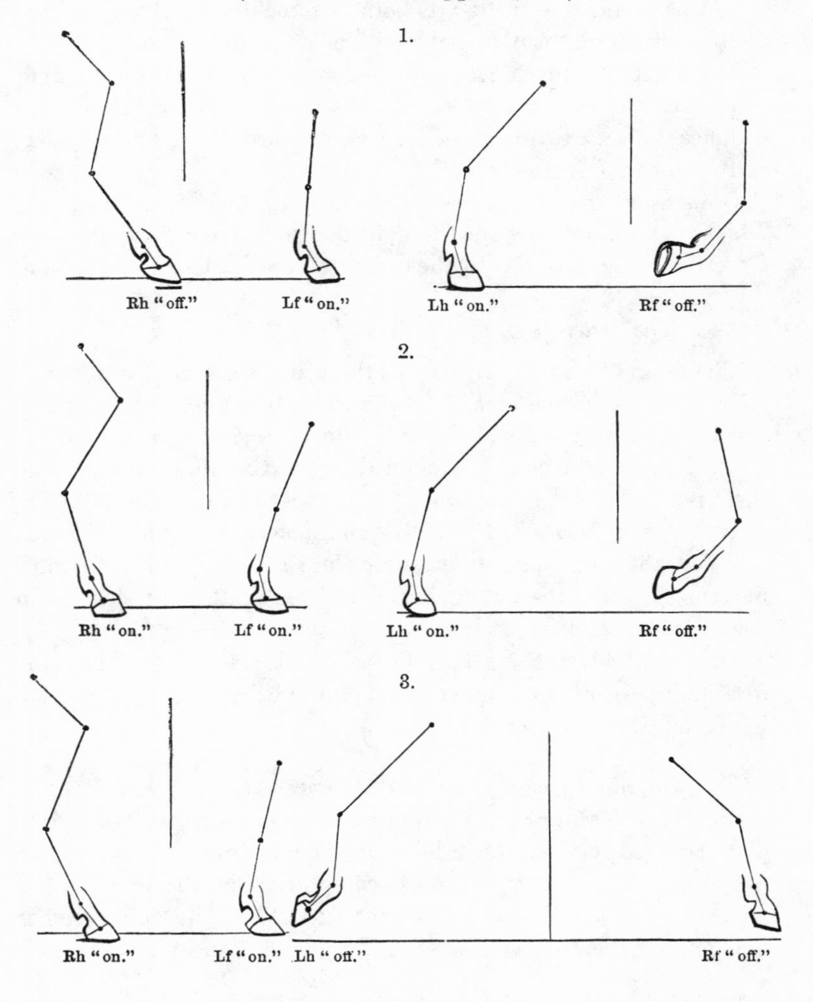

4.

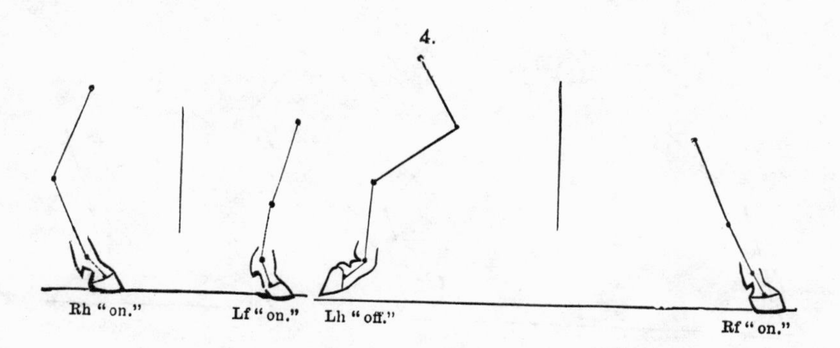
Rh "on."
Lf "on."
Lh "off."
Rf "on."

5.

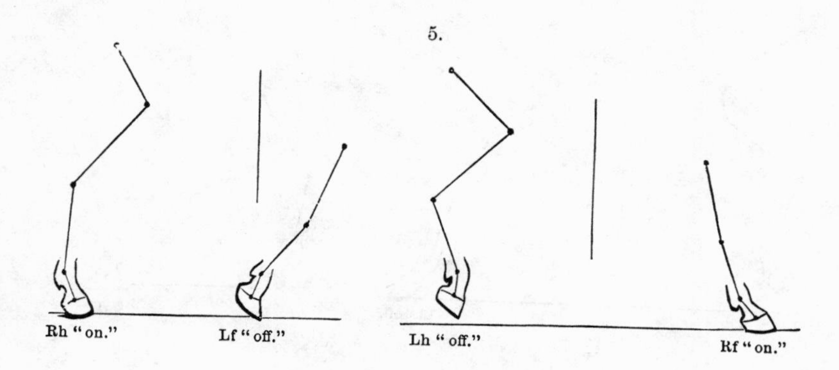
Rh "on."
Lf "off."
Lh "off."
Rf "on."

6.

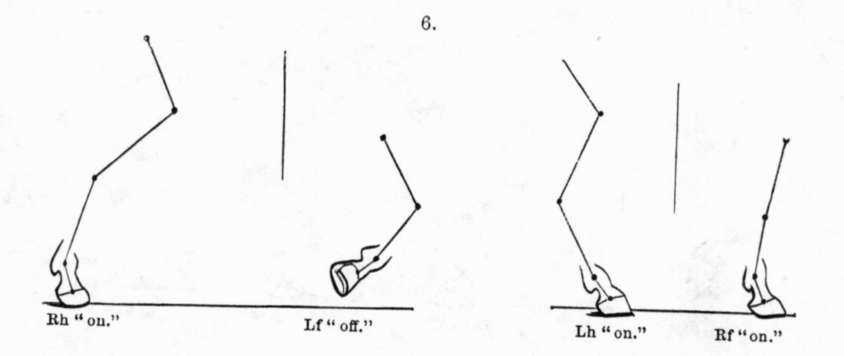
Rh "on."
Lf "off."
Lh "on."
Rf "on."

7.

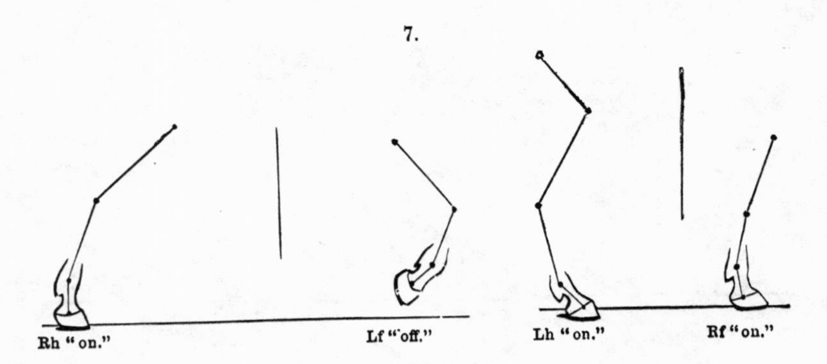
Rh "on."
Lf "off."
Lh "on."
Rf "on."

8.

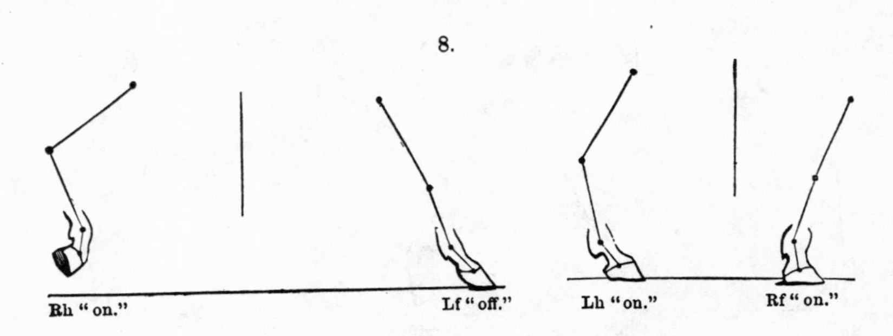
Rh "on."
Lf "off."
Lh "on."
Rf "on."

9

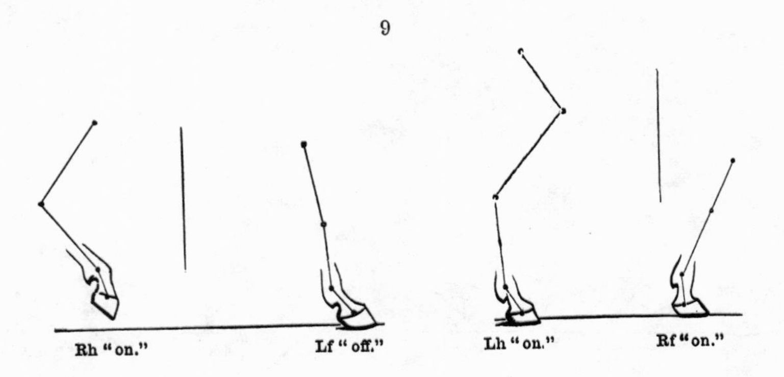
Rh "on."
Lf "off."
Lh "on."
Rf "on."

10.

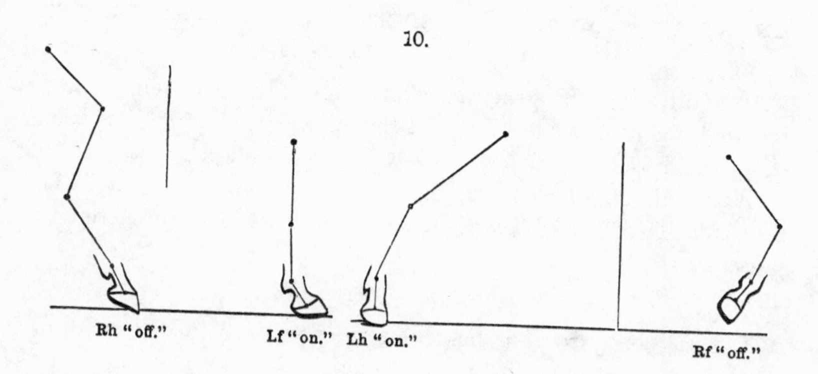
Rh "off."
Lf "on."
Lh "on."
Rf "off."

11.

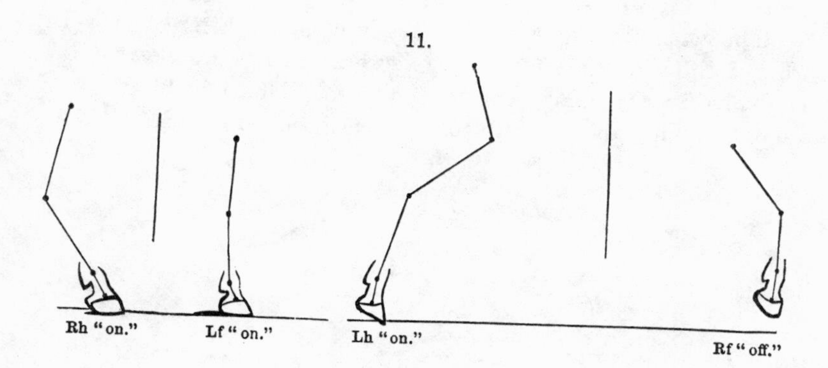
Rh "on."
Lf "on."
Lh "on."
Rf "off."

12.

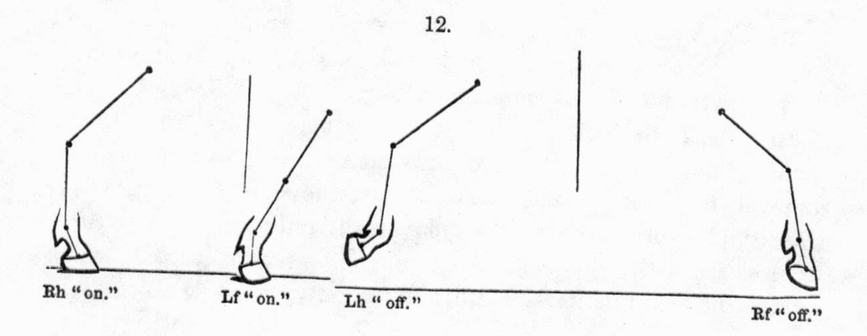
Rh "on."
Lf "on."
Lh "off."
Rf "off."

An Analysis of the Movements of the Raccoon, the Baboon, and the Sloth.

In addition to the general remarks on animal movement and the analysis of separate phases of the limb positions, it has been thought advisable to describe a few forms in detail.

With this object in view, figures of the raccoon (series 744), the baboon (series 747), and the sloth (series 750) are herewith presented.

The Raccoon. (Series 744.)

The raccoon (Procyon) has been accepted as a form worthy of careful analysis, since the genus is one of the most ancient of the extant Carnivora, and presents in its gait opportunity for ascertaining the manner in which a primitive flesh-eater moved when of small size and of short limb. (See p. 38.)

The gait from Fig. 1 to Fig. 4 is in diagonal heterochiry. In Fig. 1 the right fore foot lingers so long on the ground that the left hind foot is about to come "on" also. The right hind foot is about leaving, and the fifth toe projects backward. (See p. 51.)

(Fig. 2 of series.)

In Fig. 2 the right fore foot remains on the ground, the left hind foot now reaches the ground at its side, and the right hind foot is "off." The position is retained in Figs. 3, 4, and 5.

In Figs. 5, 6, 7 the gait is left lateral heterochiry, the support being on the left legs, the fore limb being in forward movement and the hind limb being in backward movement. Fig. 8 is a transitional form (the animal movement only resting on three legs) back to the diagonal heterochiry, as seen in Fig. 9, which retains the combination in one picture only, to be again in right lateral heterochiry in Figs. 10, 11, 12. The fifth toe of the right hind foot is here seen to extend horizontally. The change to

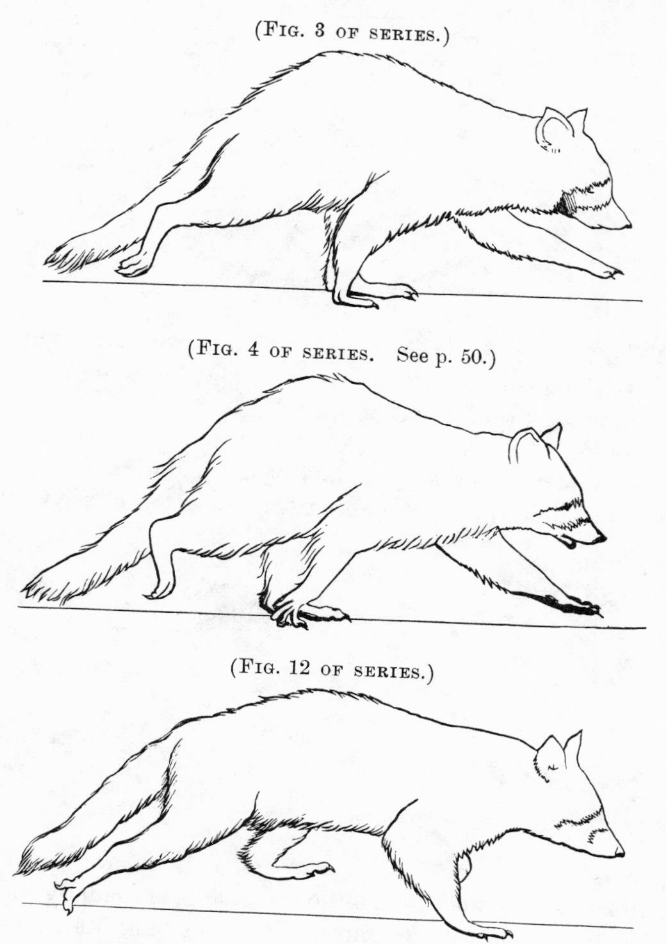

(Fig. 3 of series.)

(Fig. 4 of series. See p. 50.)

(Fig. 12 of series.)

diagonal heterochiry is made in Figs. 13, 14 (which are repetitions of Figs. 1, 2, 3), and a stride is completed. The animal now alters its gait from heterochiry to synchiry. Fig. 15 is the form of transition. In Fig. 16 the change has been completed and the animal is sustained alone by both hind feet. He maintains this position in Fig. 17, but again is in a transition form of three legs on the ground in Fig. 18.

In Fig. 19 the animal is in diagonal heterochiry, in which position the right hind leg shows the fifth toe in the axis of the leg, and greatly abducted from the fourth.

(FIG. 19 OF SERIES.)

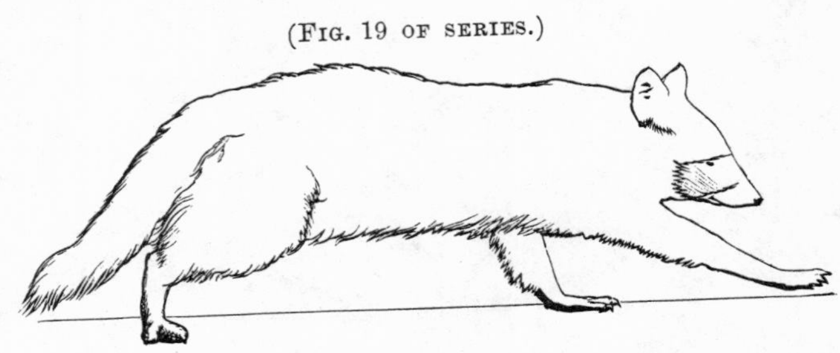

In Fig. 20 it is again synchiral, the fore pair now assuming the work of supporting the body. The fore limbs are nearly at the extremes of forward and backward movement.

The sole of the foot in the left fore foot is in part in aid of the limb, and presents an example of the plantigrade foot becoming semi-digitigrade as the foot passes backward of the vertical line.

(FIG. 20 OF SERIES.)

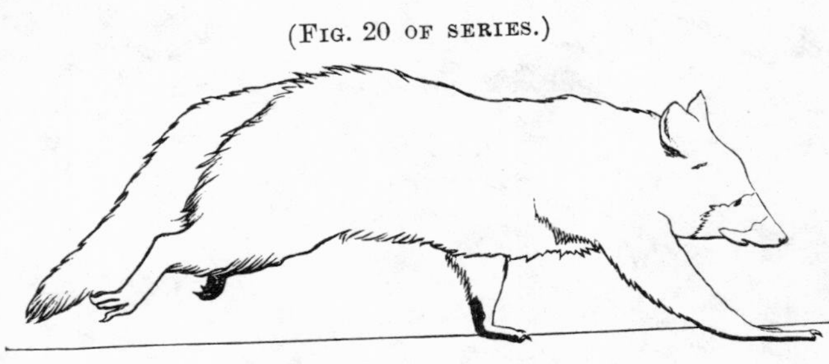

It will be seen that the gait is irregular; it is first a walk or trot, afterwards assumes the position of pacing, and finally ends in a gallop, or as much of the latter gait as can be assumed to exist in an animal with short legs.

The Baboon. (Series 747.)

In the series of the baboon the animal is seen in the walk. The form was accepted for comparison with the raccoon because of the position of the animal near the other end of the scale of quadrupeds. The limbs have great length, and the intervals between the two feet of a single pair when in extremes of forward and backward movement are correspondingly great.

In Fig. 1 the animal has just left lateral support, in which both limbs move in the centre of the body by the left fore foot leaving the ground. In Fig. 2 the support is diagonal, and both limbs (right fore and left hind) are removed from the centre. This is retained through Figs. 3 and 4. In Fig. 5 the same dis-

(FIG. 5 OF SERIES.)

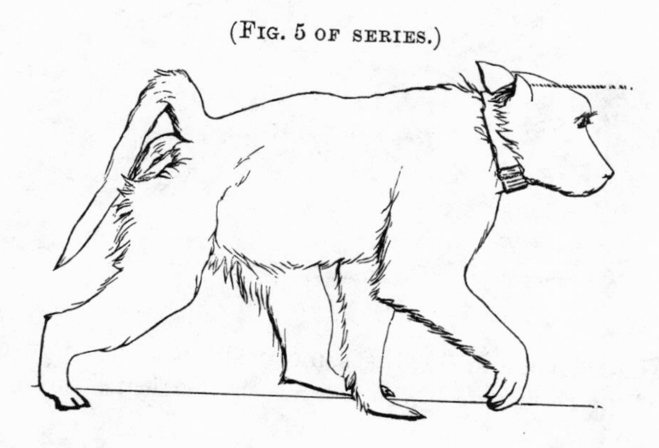

position continues, but it is noteworthy because of the shape of the left hind limb, which resembles the shape of the ungulate foot. The right fore limb is now in the extreme of backward movement.

(FIG. 6 OF SERIES.)

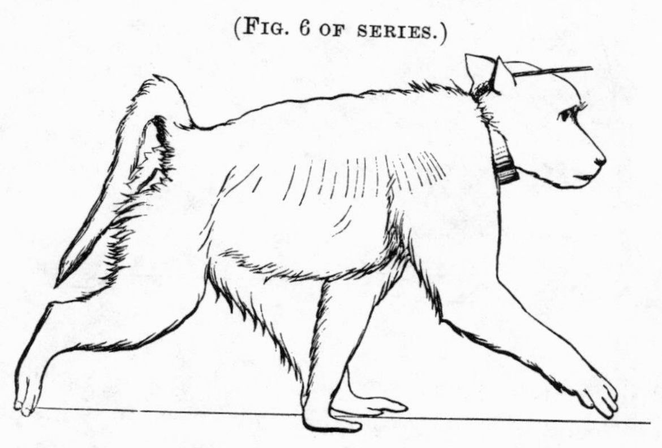

In Figs. 6 and 7 the position is left fore and right hind (lateral heterochiry).

In Fig. 8 the laterals have changed to diagonals by the right fore foot coming down as the left leaves the ground.

(FIG. 7 OF SERIES.)

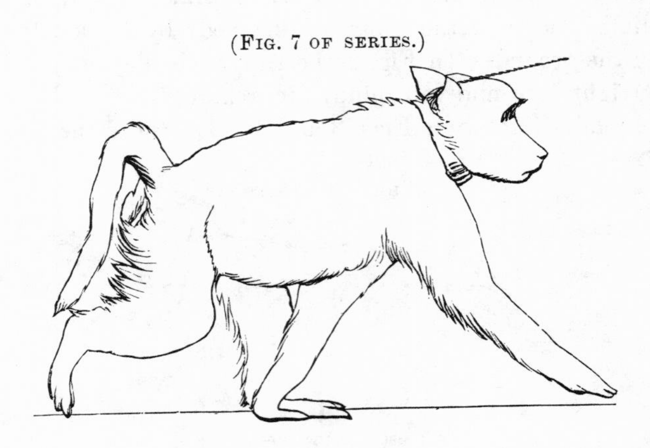

In Figs. 9, 10, and 11 the same position is maintained. The shape of the right fore leg as it nears the vertical position is noteworthy.

(FIG. 8 OF SERIES.)

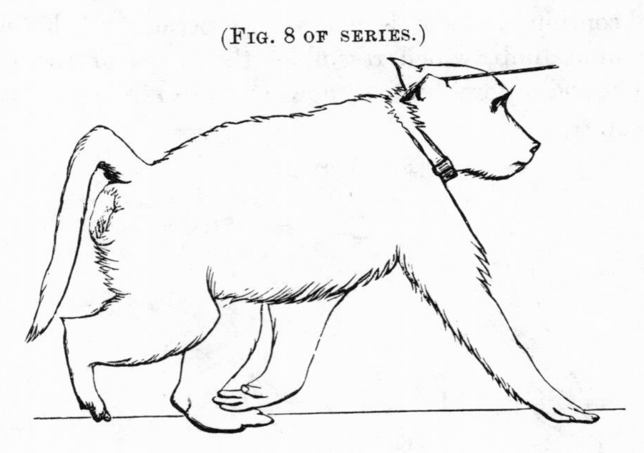

The left hind foot has overtaken the lingering right fore foot, and three feet are on the ground. This position is instantly lost by the lifting of the left hind foot from the ground, when a diagonal support ensues, which closely resembles in general effect the lateral support in Fig. 6.

(Fig. 11 of series.)

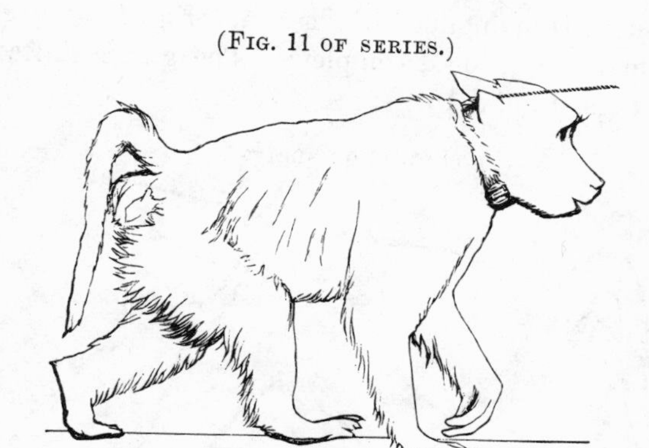

(Fig. 12 of series.)

(Fig. 13 of series.)

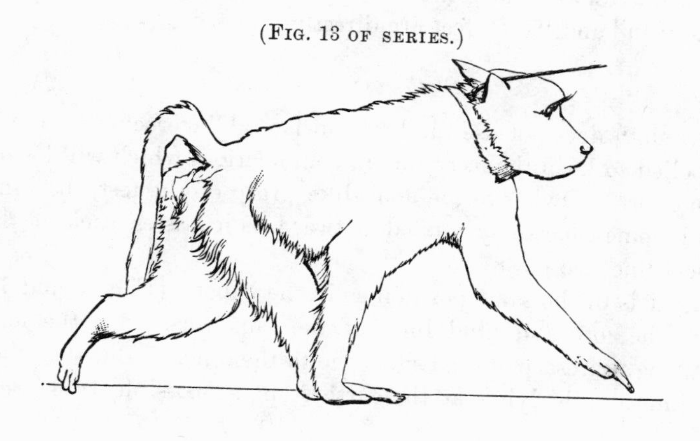

This position is maintained in Fig. 14. Fig. 14 is a repetition of Fig. 1, and the stride is complete. The gait is shifted to left laterals in Figs. 15 and 16.

(FIG. 14 OF SERIES.)

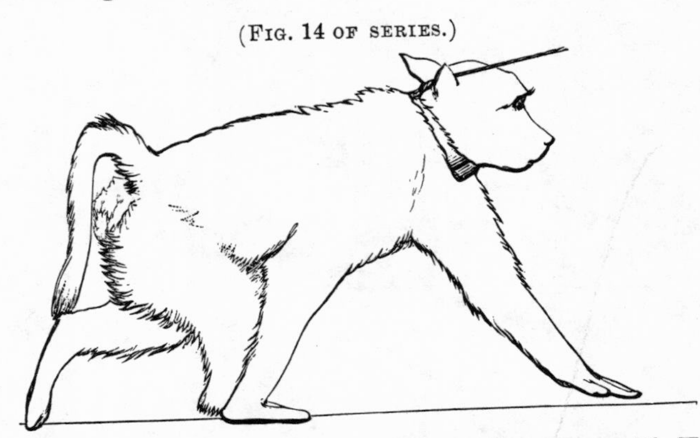

Figs. 13 and 14 can be compared with advantage with Figs. 3 and 4, 13 and 14 of Procyon (series 744). The feet are relatively in the same position.

In the baboon the dorsal line changes scarcely at all from that in the other figures. In Procyon the dorsal line is more convex in these figures than in any others.

The most characteristic position of Procyon is the laterals in Figs. 11 and 12. The limbs are in the extreme of movement, the hind backward and the fore forward. The most characteristic position of the baboon (Fig. 6) is the reverse of this, the hind limb being forward and the fore backward. The relative places of the "on" and "off" feet are directly opposed.

The Sloth. (Series 750.)

The sloth does not use the limbs in lateral heterochiry. He is compelled to keep the feet "on" in combinations which will insure his support. Owing to the hook-like grasp of the feet, the hind limb becomes markedly rotated outward as it passes back of the vertical line (see p. 38).

Fig. 1 is in the same position as in the baboon (Figs. 6 and 7), —*i.e.*, the fore and hind limb of the same size (viz., the left) are at the nearest point possible one to the other. But the other two limbs, while lying at the farthest point possible from each

(FIG. 1 OF SERIES.)

(FIG. 2 OF SERIES.)

(FIG. 3 OF SERIES.)

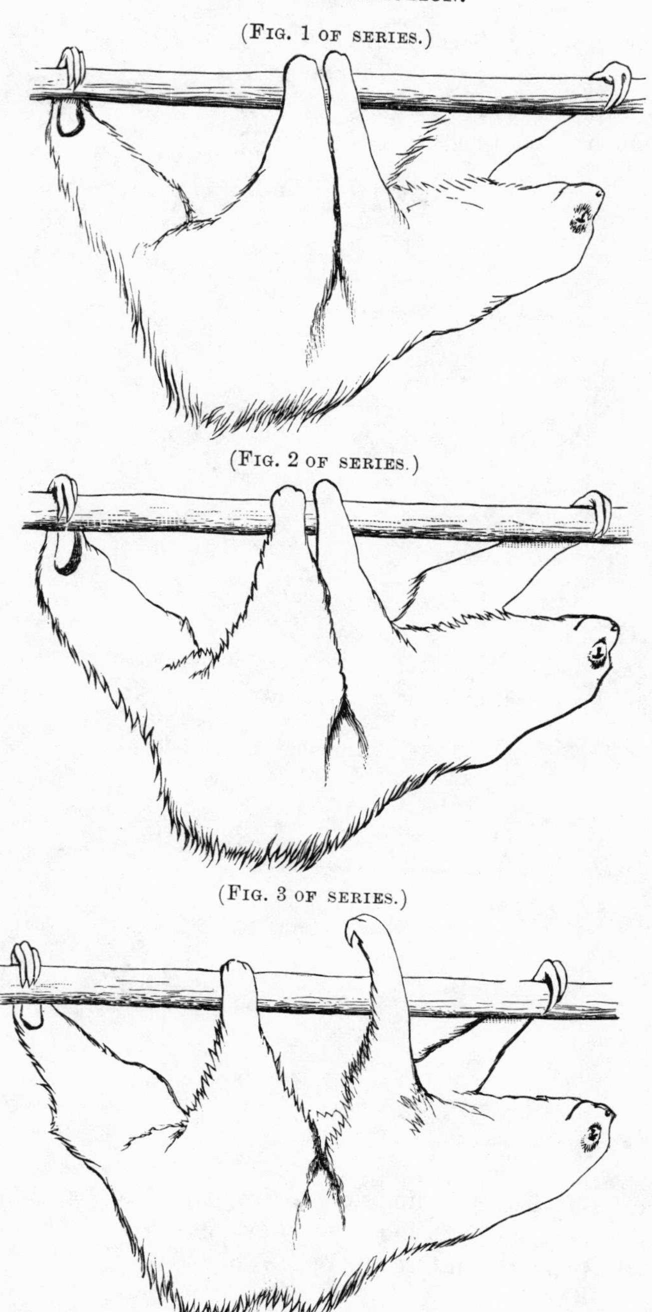

other, are not "off," as is the case with the baboon. Save in this instance, the positions are the same.

The attitude remains unchanged in Fig. 2. In Fig. 3 the left fore limb has left the perch, and is projected directly upward.

(FIG. 4 OF SERIES.)

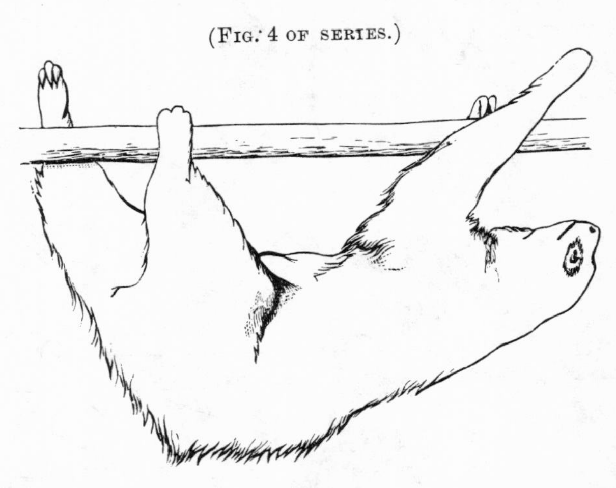

In Figs. 4 and 5 the left fore limb is swung forward above the level of the perch, and the right hind leg is "off" and projects

(FIG. 5 OF SERIES.)

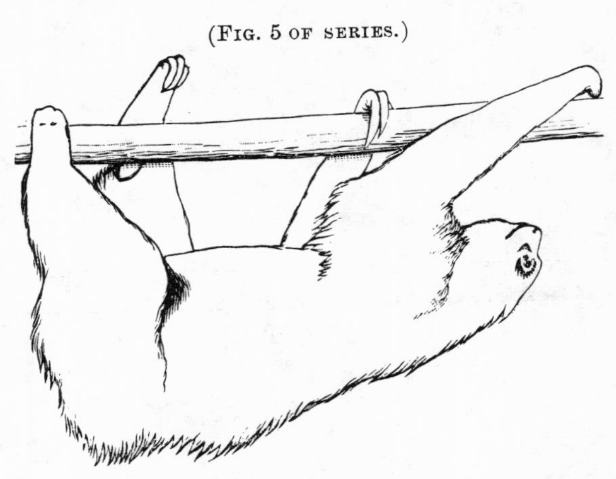

upward. In Fig. 6 all limbs are again "on," as in Fig. 1. The left fore and the right hind limb have just come "on." The position is maintained through Figs. 7, 8, and 9.

(FIG. 6 OF SERIES.)

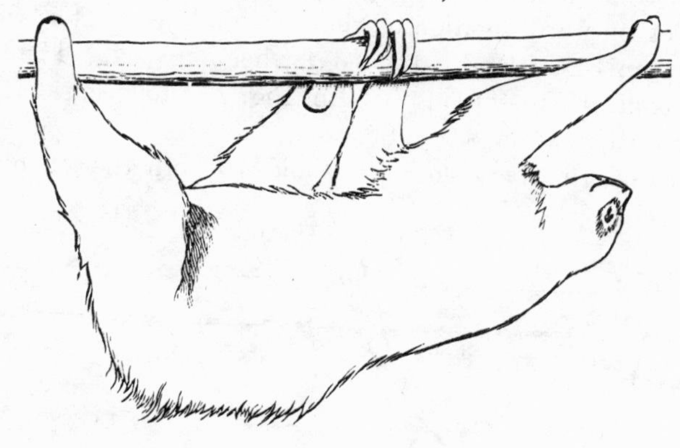

(FIG. 7 OF SERIES.)

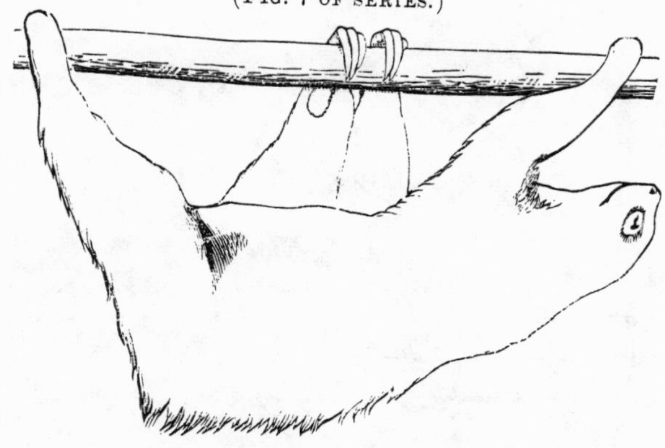

(FIG. 8 OF SERIES.)

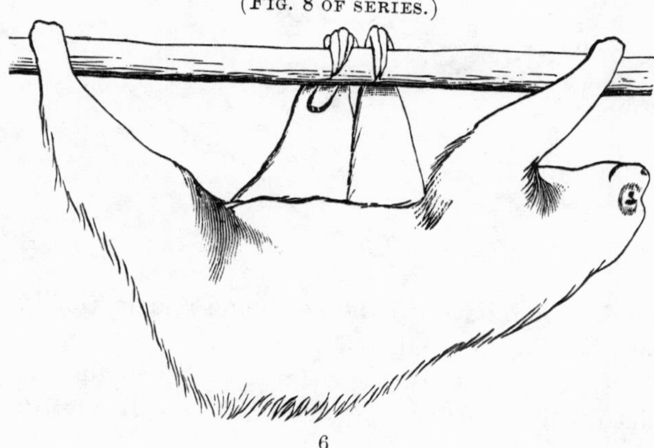

Figs. 7 and 8 closely resemble each other. They differ in the degree of outward rotation of the left hind leg. It is wider in Fig. 8 than Fig. 7, and is rotated farther outward. The head in Fig. 8 is nearer the beam than in Fig. 7. For consideration of the outward rotation, see p. 89.

In the interval between Figs. 9 and 10 both the left hind and

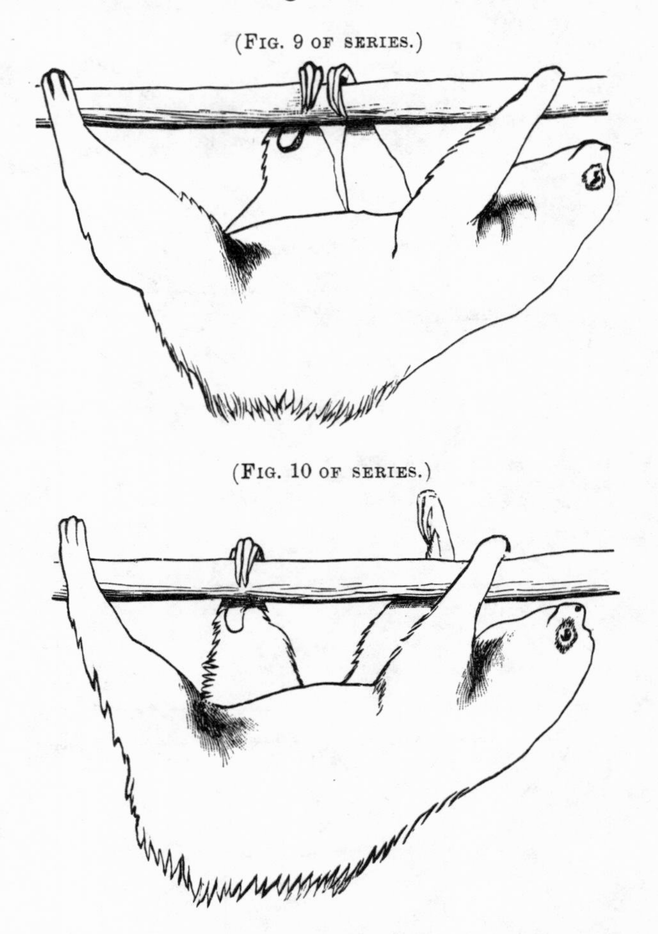

(FIG. 9 OF SERIES.)

(FIG. 10 OF SERIES.)

the right fore limb have left the perch and remain "off" through the remaining Figs. 11 and 12.

In Figs. 10, 11, and 12 the manner in which the support is maintained while the two limbs are "off" is illustrated. It is strictly diagonal heterochiry. Yet in Fig. 12 the same positions as in Figs. 1, 6, 7, 8, and 9 are about to be assumed, showing that,

(FIG. 11 OF SERIES.)

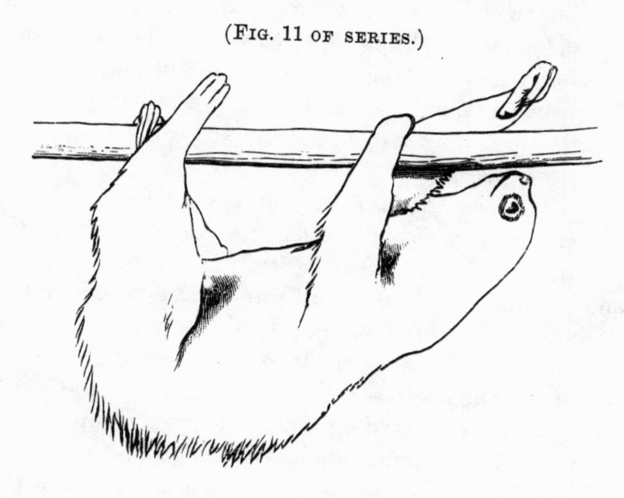

while support may be diagonal, the limbs which relatively are farthest apart or the nearest together are laterals. Thus the gait

(FIG. 12 OF SERIES.)

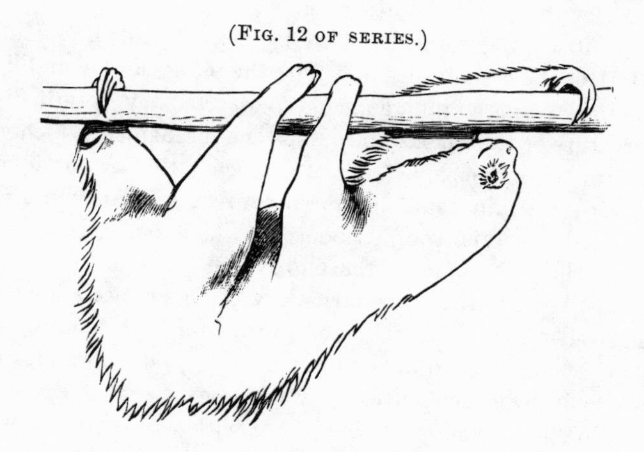

of the sloth is peculiar in having all four feet "on" at one stage of movement, while the gait is the usual disposition of alternate feet by diagonal.

ZOOLOGICAL CONSIDERATIONS.

The object which the writer has had constantly in view in studying the photographs is to endeavor to determine the value of attitude and gait to classification. Many of the notes bear to a greater or less extent upon zoology, but a few which appear to apply in an especial way to taxonomy are conveniently brought together at this place.

As is well known, the manus and pes present a plan in the arrangement of the elements which may conveniently receive the name of the lateral and median series. In the manus, the cuneiform and the unciform bones constitute the lateral, and all the remaining bones the median, series. In the pes, the calcaneum and cuboid bone are embraced in the lateral series, and all the remaining bones form the median. With this necessary understanding, the writer will now proceed to state the premise upon which he undertook a special examination of the manus and pes in connection with the phenomena of limb-movement. The episode may be accepted as an instance in which the photographs suggest various lines of research, even though nothing of especial value be claimed for the results here secured.

When it was determined, chiefly by the study of the stroke (see p. 50), that the foot comes down by the outer border and leaves by the inner, an attempt was made to ascertain by examination of the structure of the carpus and tarsus the mechanism which correlated with the movement. It was expected that an oblique line by which the strain could be traced across the small bones would be found. The views of Leboucq,* which embrace a path of precisely this character in the embryonic form of the tarsus of mammals, appeared to be confirmatory of this expectation.

The well-known fact that the calcaneum occasionally ossifies with the scaphoid bone in the human foot,† and is at times found united to it by synchondrosis,‡ was suggestive that the oblique axis could be determined in that form with whose structure anat-

---

* Arch. Biol. (Brussels), iv. 35.

† Gruber, Mém. de l'Acad. de St. Petersbourg, ser. vii., p. 9, Taf. xvii.; Holl, Arch. für Klin. Chirurg., 1880; Zuckerkandl, Wiener Med. Jahrbuch, 1880.

‡ Weber M., Verslag. der kon. Ak. Amsterdam, 1883, xviii. 121.

omists are most familiar. Aeby* describes a similar arrangement in the foot of the gorilla.

It was reasonable to infer from these data that the disposition of the calcaneum to transfer the strain transmitted to it across the foot to the scaphoid would characterize the mammalian foot.

A careful examination of the *tarsus* has not warranted the inference. As shown by Leboucq, the arrangement exists in the embryo, but it is not maintained as a rule in the adult, nor is it seen in the phylogeny of the group.

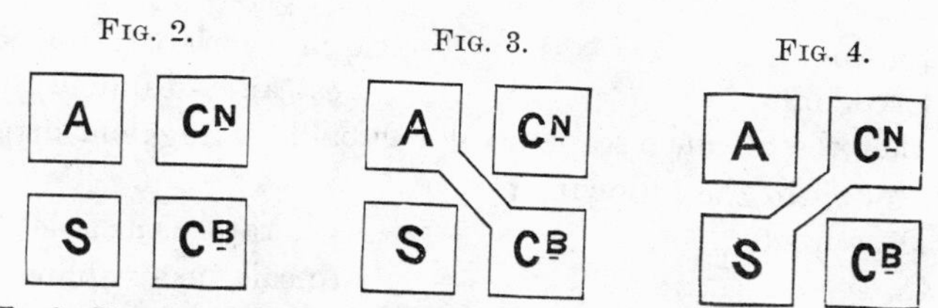

Fig. 2.—Diagram of the relation of the astragalus (A), the calcaneum (Cn), the scaphoid bone (S), and the cuboid bone (Cb) in the foot of mammals.

Fig. 3.—Diagram to exhibit the disposition for the astragalus to form a junction with the cuboid bone. (Letters as in Fig. 2.)

Fig. 4.—Diagram to exhibit a disposition for the calcaneum to form a junction with the scaphoid bone. (Letters as in Fig. 2.)

The statement of Kowalevnsky is accepted, that the earlier forms of mammals exhibit lines of support which correspond to the arrangement of the bones of the leg. (See Fig. 2.) But in the animals now living, the lines of the bones of the carpus and tarsus do not as a rule show such independence. The bones of the tarsus are almost invariably more or less displaced. In the tapir and the rhinoceros, which are the best living examples of related forms to the earlier forms, the acquired support from adjusting rows is as exact as in any of the numerous recent specialized forms which at present inhabit the earth.

So far as examined, the calcaneum articulates with the scaphoid bone in the following genera only:

Homo,
Troglodytes,
Equus (variable; sometimes astragalo-cuboid),
Dipus,
Cœlogenys,
Erethizon (variable).

* Morph. Jahrbuch, iv., 1878.

Herewith are given examples of the common cuboido-astragaloid articulation:

Hyrax (not marked),
Rhinoceros,
Amynodon,
Ungulata,
Procyon,
Ursus,
Erignathus,
Menodus,
Palæosyops,
Creodonta,
Macropus,
Phascalarctos,
Dicotyles,
Hippopotamus,
Erethizon (variable),
Arctomys,
Myrmecophaga,
Elephas,
Equus (variable; sometimes calcaneo-scaphoid).

Neither the calcaneo-scaphoid nor cuboido-astragaloid articulation is found in the following:

Dasyprocta,
Capromys,
Fiber,
Hystrix,
Hyena crocuta,
Orycteropus (variable; tends to calcaneo-scaphoid),
Homo (variable; sometimes calcaneo-scaphoid).

While a fixed arrangement of the parts in an assumed oblique axis is thus undetected, the examination was fruitful of one result,—namely, that when in addition to the axial disposition of the tarsal bones one to another and the lateral adaptation of parts in transverse rows was departed from, the departure took place in one of two ways,—either by the calcaneum reaching the scaphoid (*i.e.*, by a direction downward and inward) or by the cuboid bone reaching the astragalus (*i.e.*, by a direction upward and inward). It is thus demonstrable that the elements in the tarsus which first distribute the weight of the outer border of the foot as it comes to the ground transmit the impact *inward* in an *oblique* manner, and, it may be assumed, tend to throw the weight of the limb towards its axial line, which sustains the weight of the body when the limb is in the vertical position.

In the study of the phylogeny it will be seen that the more distal of the two bones is first pressed inward, and that this arrangement obtained not only in the first (with the exception of Coryphodon), but in most of the extant forms. Why it should ever have departed from this arrangement and reversed the inward impact to the calcaneum is not an easy question to answer.

It is worthy of remark that in Equus and Dipus we have highly specialized forms of foot-structures; and in Primates, while the foot on the whole is generalized, the astragalus sends a remarkably inward directed and obliquely placed neck in a direction which would render it difficult for the cuboid bone to reach it.

It must be acknowledged as a great difficulty how an intermediate group arose in which a simple form of tarsal disposition is retained. While from facts presented by Professor Flower it would be natural to assume that Orycteropus might move the foot in a manner different from Myrmecophaga, nothing can be urged why Hyena should differ from Canis, or Dasyprocta and Capromys from Arctomys.

An examination of the *carpus* showed less disposition for the outer elements of the series to be placed inward than is the case in the tarsus. The cuneiform bone, which corresponds in position to the calcaneum, and the unciform, which corresponds to the cuboid bone, remain in all forms of the manus in constant relation to the rest of the carpus. But the entire lateral division of the carpus exerts a disposition to displace the os magnus inward and upward in a manner which is strikingly exhibited in the following scheme.

The os magnum articulates with the scaphoid bone in

Cœlogenys,
Rhinoceros,
Dicotyles,
Hippopotamus,
Tapirus,
Sus,
Chrysochloris.

That is to say, the os magnum, in addition to articulating with its own axially disposed element, the lunare, secures an articulation with the scaphoid bone.

The os magnum articulates with the lunare only in

Dasyprocta,
Capromys,
Erethizon,
Hyrax,
Hystrix,
Arctomys,
Uintatherium,
Elephas (variable with trapezoidal-lunare),
Tapirus,
Manatus,
Orycteropus,
Myrmecophaga,
Myogale.

In summarizing these results, it may be said that the fact of the outer border of the pes coming to the ground in the first stage of the stroke is in correlation with a disposition for one or more of the tarsal elements of the outer series to secure junctures—*i.e.*, junctures not accounted for in the plan of the parts—with one or more elements of the median series. Relatively few genera resist this disposition.

In the manus, the lateral series remains intact, but in a few extant genera and in some extinct genera the outermost of the median series is seen to articulate with a carpal element which lies to its inner side.

Deviations of the Trunk with Respect to a Hypothetical Longitudinal Axis.

An interesting series of comparisons can be instituted between the anterior and the posterior parts of the trunk in the various positions of the limbs. In illustration of this subject attention is directed to the deviations of the trunk in advance of the rump with respect to a line drawn horizontally forward from the base of the tail. Assuming that the beginning of such a line corresponds to the part which shows the least motion and passes to that which shows the most, it will follow that the fore part of the trunk, together with the whole of the regions of the head and the neck, will shift positions with respect to this line. In series 675 the figures of the hog demonstrate that the body in advance of the rump are highest when the animal is resting on both hind feet. It is next highest when the animal has come down one fore foot. In the deer, when propped on the fore limbs, the animal sinks so low anteriorly as to permit the horizontal line to intersect the muzzle. The fact that the baboon exhibits no contrast of the kind named would lead to the conclusion that the clavicle prevents the fore part of the trunk from sinking,—a conclusion which is invalidated by the fact that the elephant shows a similar exemption.

Relation between the Oblique Movement across the Foot and the Number of Toes.

From the fact that the outer border of the foot first strikes the ground and that the inner border last leaves it, one can create a proposition to which the following facts are corollary. (See pp. 42, 50.)

When the outer border is longer and stouter than the inner, the first-named parts of support are of more importance than the last. The difference in this regard between the foot of man and that of the bear should be the difference between the respective lengths of the first and the fifth toes.

When the inner toe is rudimental, or lost, the foot has left the ground before the inner border is reached. But it is not so evident why the fifth toe should be lost. One would suppose that this toe would be the most persistent. The sequence of the limbs of the horse from the five- to the single-toed forms would be of interest to study in this connection.

Rotation of the Limb.

In another relation than that included in the foregoing the study of the photographs has been of interest. Allusion is particularly made to the forward swing of the fore limb and the outward rotation of both limbs (for inward rotation see p. 59).

The extent of the forward swing is found to correspond to the extent of the trochlear surface of the humerus as it enters into the composition of the elbow-joint. In addition to this fact attention may be called to the difference in the length of the olecranon in different animals. The olecranon is greatly longer relatively to the shaft of the ulna in primitive types of Mammalia than in the more specialized, as the writer has been informed by Professor W. B. Scott. The same difference is found to obtain between Dicotyles and Sus. May it not be claimed that the difference correlates with the facilities possessed of maintaining the fore limb in backward strain? The enormous leverage secured by a long olecranon must enable the limb to maintain a strong position for the limb when it is held in backward strain as compared to one in which the olecranon is short. This must be especially marked when the humerus is long, as is always the case in these types.

The mechanism which accompanies the rotation of the limb outward at the knee after it has passed the vertical line is associated, as has been seen, with inward rotation of the hock. This form of torsion is described (see p. 45), and the springing power of the limb is a resultant. The amount of springing power should correlate with the torsion, and the facets by which twists are effected be indices of the act. The conversion of the calcaneocuboidal articulation into a convexo-concave joint, in forms in which the oblique impact is cuboido-astragaloid, would look as though such impact favored torsion.

The study of the rotation of the limb is conveniently undertaken in the elephant, since the fur is here absent. A number of the positions assumed by the limbs is illustrated in series 733, Figs. 1 to 12. (See the figures of the fore limb of the elephant.)

The width of the limb at the elbow is observed to be less in backward strain than in the recover.

In Figs. 1, 2, 3, and 4 the olecranon is in line with the posterior border of the arm above the elbow.

In Fig. 4 the foot is leaving by its inner border. In Fig. 5 the olecranon is more in view, since it is turned out, at the same time the foot is now everted (*i.e.*, rotation outward has taken place as the limb passes from impact of backward strain to that of recover). Extreme position of eversion (Fig. 8) is coincident with the distance between the olecranon and the humerus, and the greatest reduction of distance between humerus and anterior border of the arm. The posterior border of the arm is here beginning to leave the side of the trunk, and a white light is reflected from over the triceps muscle.

During the forward movements of Figs. 9 to 12 the limb is slowly passing back to the position where the olecranon is again in line with the posterior border of the limb (*i.e.*, strictly speaking, in profile of the limb), and the foot is preparing for a second period of impact against the ground.

The lighter shade of the prominence at the back of the elbow answers to the position of the olecranon; the elongated mass of light shade which extends parallel to the arm between the elbow and the shoulder answers to the position of the outer border of the humerus. It is evident by the pictures that the distance between the light over the humerus and over the olecranon is variable. But it is known that the actual distance is invariable, and therefore the apparent variance is due to the motion of the limb in rotation at the shoulder, by which the region of the back of the elbow is moved somewhat outward. At the same time the distance from the outer border of the humerus and the anterior contour is lessened as rotation takes place. These facts being borne in mind the following additional observations can be made:

The torsion of the limb is pronounced in the Tylopoda examined,—viz., the camel, the dromedary, and the guanaco. This fact should not be disassociated with the absence of anchylosis between the cuboid bone and scaphoid bone with these animals.

In contrasting the above examples with the Pecora,—*e.g.*, the other ruminants, excepting the family of the musk-deers, not examined,—the following structural features may be mentioned as correlations of peculiarities of gait.

In the Tylopoda, when the limb is supporting the body, the hoof is always in line of support with the rest of the limb. In the deer the hoof and digits are angulated with the rest of the limb.

In the Tylopoda the joints are thick, and the outlines are distinctly seen. The fore limb of the Tylopoda is never noticed at a point posterior to the withers, but leaves the ground at an angle of about minus forty degrees. The fore limb of the Pecora is vertical at a point at the centre of the trunk, and leaves the ground at about minus fifteen to twenty degrees. When the limb is off the ground, the digits are flexed to a greater degree in the Tylopoda than in the Pecora.

In the Pecora the limb is smaller and more delicate in proportion to the bulk of the body; the spread of the hoof is less, and the movement of the bones each in the other more precise than in the Tylopoda.

The vertical position of the fore limb under the centre of the trunk is possible, because of the oblique position of the humerus. The heavily-built bovine animals present the same contrasts as the deer, with the exception that the fore limb is carried less backward. (See p. 45.)

The camel has a relatively short metapodium. The joints present curved contours in place of angular ones,—excepting the angle formed by the hock,—and even this is less pronounced than in other ruminants.

Minor Peculiarities.

While the general features of the quadrupedal motion are in all animals the same, minor peculiarities of the carriage have doubtless taxonomic significance. A number of these will be indicated.

The dog when off the ground flexes the fore legs, the cat extends them. This probably correlates with the size of the supraspinous fossæ of the scapulæ in the two animals. (See p. 84.)

The extended fore leg in the cat is parallel to the long axis of the trunk, while in all the ungulates it forms an angle to this axis. The angulation of the parts in ungulates is subject to variation, being more open in the deer tribe than in the ox; in a word, in the lighter-bodied as contrasted with the heavy-bodied forms. The humerus in backward movement is nearly horizontal in the Pecora. (See pp. 42, 45.)

The enormous size of the anterior dorsal spines in the hoofed animals, their smaller size in the claw-footed, their entire absence in Dipus and the bat (in which animals the fore legs are not used for support), the trend of the great extrinsic muscles of the fore limb towards the head, all prepare the observer for some taxonomic values in the relations existing between the strength of the cephalo-humeral muscles, the obliquity of the fore leg in forward movement, and the size of the anterior dorsal spines, and possibly in the shapes of the atlas and axis.

The camel protrudes and withdraws the under lip each time the fore and hind limbs are shifted.

In the North American buffalo the prepuce is drawn backward when the hind limbs are in backward strain.

Characters pertaining to the Superficies emphasized in the Photographs.

It is a well-known circumstance that the photograph will bring out details of structure which are not seen by the unaided eye. Thus, an inscription on a tombstone which was covered by lichen and was quite illegible was found by Professor J. T. Rothrock to be read with the greatest ease when seen photographed after the ordinary manner. The photographs of the series of animals studied reveal similar interesting features. The spots on the lion and the lioness are much more distinct than they are in the living models. In the lioness, in addition, five minute points are detected on the back of the trunk. They probably correspond to the eminences of the skin over the spines of the dorsal vertebræ.

A fairly recognizable description could be drawn up of the superficial muscles of the limbs, especially of those fleshy masses which pass from the trunk to the limbs. The reflections of light from the hair of the figures of the horse may be ascribed to the changes in degrees of convexity or concavity of the superficial muscle-masses, while some of the stripes of the tiger, especially those of the fore limb and the neck, will be found to answer to the depressions existing between well-defined positions of muscles which can be named with accuracy. In the figures of the roan horse (series 582) the black patch on the rump corresponds to the great biceps muscle.

In the horse (series 579) the depressions between the muscles of the posterior aspect of the thigh are most marked when the limb is in the first stage of the recover, but are obliterated when the limb is in backward strain. The masses at the inner border of the thigh, as seen in the horse (series 582), are nearly flat when the limb is off the ground, but conspicuously convex when the animal is using the limb for support.

In the dog, when both fore feet are off the ground, the muscles which pass from the trunk to the fore limbs are exceedingly tense.

On the Significance of the Skin-Folds on the Trunk of the Hog.

A number of minute changes in the contour of the animals are worth noting. In the figures of the hog (series 673 to 675), at the time when the limbs of a single side are the nearest one to the other the skin is observed to be thrown into a series of vertical folds. It is suggestive of the mechanical origin of the bands in the armadillo that the lines in the hog are in the same position as the bands. They appear to differ only in the circumstance that the folds are transient, while the bands are permanent. (See p. 45.)

The Mane.

Each time the horse, in the series 590, leaves the ground by one of the hind feet the mane streams backward. In series 638, as the animal descends the mane streams upward. This change in position is caused by the motions of the head and neck upward and downward. (See especially figures of the horse rocking, series 649 A. See p. 48.)

The Significance of the Keel on the Metapodium.

The presence of a ridge on the distal ends of the bones of the tarsus and carpus, as well as on the first and second rows of phalanges, is indicative of a precision of motion in the joints of which it constitutes a part. Such ridges are known as keels, and are found in the feet of the ruminants, excepting the group of the camels, and in the horse. The primitive types of mammalian life, as has been pointed out by Professor E. D. Cope,* were without the keel. It first appeared on the posterior surface of the articular surface, and subsequently upon its anterior surface as well. These facts suggest the conclusion that the ridge was developed for some exigencies of the foot while in backward strain, and in the action of the transition from "on" to "off," since the keel first appears on the flexor side of the foot, and that, since in extant animals the camel and its kin are without the keel, a careful comparison of the motions in these creatures with those of the other ruminants would be suggestive of the use of the keel, and lead to some interesting differences between the gaits of primitive and specialized types.

After having made careful studies of the series of the Egyptian camel (736), the Bactrian camel (737 to 742), and the guanaco (743), I have been unable to find any features which correlate with these anatomical peculiarities.

Evolution of the Gait.

The gait in its various expressions is based upon the correlations between the fore and the hind limbs. Assuming that the simplest movement is an alternate action of a single pair (synchiry), the simplest gait is that in which this alternation occurs in each pair independently. The gallop is thus the simplest gait, since the fore pair and the hind pair are independent. But the gallop depends upon a high momentum, and we must assume that the primitive animals were slow walkers. When in a slow rate the same independence is observed,—*i.e.*, when the right fore foot is moved forward at the same time as the right hind foot (lateral heterochiry),—the two pair act rhythmically, and the feet "keep step." This is seen in the rack, which thus becomes a

---

* The Origin of the Fittest.

practicable gait for a primitive creature. The only objection to this conclusion is the fact that in the giraffe, which is a specialized and aberrant form,—and the legs are remarkable for their length,—we have the best example of a natural pace.

The simplest forms of moving appendages are seen in the Nerita, one of the fresh-water worms, in which the bristle-like rods are moved in all directions without concert. In other forms, as in the sea-hare (Aphrodite), the bristles move in numerous lines extending alongside of the body, but each line in turn being independent of those adjacent to it. The next arrangement obtains in myriapods, whose feet, while in pairs, are seen to move as the horse's in pacing,—*i.e.*, each row acting as a whole, and the two rows alternating. In the lower crustaceans the same method obtains. In the higher crustaceans, as in the crab, the two parts of a single pair may act together.

It stands to reason that when to one side of the body a reinforcement is received from a limb of the opposed side and of a different pair advancing to the median line, a relatively high degree of development has been attained. In a phrase, in animals with highly developed central nervous systems the bilateral coordinations become as practicable as the unilateral.

The question of the manner of using the feet in progression is also likely to be a subordinate one to the question of the size and the bulk of the body. In short-legged, heavy animals of slow movement, the support of the centre of the heavy trunk by the diagonal use of a fore and a hind limb was probably early obtained. Nevertheless the Echidna, a good example of an animal possessing a large, heavy body and short limbs, walked by diagonal heterochiry.* The feet, however, did not approach each other towards the centre of the trunk, as is the case in the walk and the trot of the horse, but kept to the side of the body.

In long, slender natatorial forms, which would occasionally slide among the ooze of a river-bed,—such an animal, for example, as the Potomogale,—doubtless specialization can be attained in the line of development of many of the gaits.† We have

---

* This was observed in the specimen at the Zoological Garden, Philadelphia.

† The crocodile of the Jumna, according to Hornaday ("Three Years in the Jungle," 1885, p. 55), can stand upon the legs in the same manner as a terrestrial quadruped.

seen that the giraffe uses the pace for all rates of movement; also that the horse modifies his paces by the character of the ground he traverses. It is probable in like manner that an animal will by natural selection determine his methods of progression. Marsh-dwellers are uniformly of the heavy-bodied, short-legged type. In an animal emerging from the marsh to the plains the limbs would become more compact and longer and the gait more various.* Caton† states that the Virginia deer bounds up once or twice at the beginning of his gait, to observe the surrounding country before he settles down to a steady run. This habit would be of no use in an animal living on a prairie, and is absent in the mule-deer. Mere speed is not a necessity of a change of gait. A trot can be based on the walk; the fast pace on the slow pace. In the opinion of Dr. M. H. Cryer, a horse in going over a hurdle employs the same division of his stride as in the run.

Variability of gait in a single animal may be accepted as an evidence of high development. Diversity of function in the study of the gait is of the same value as in other subjects of biology. Thus, an animal starting a movement on the laterals by bringing into use a member of the opposite side, so that three feet may be on the ground at the same time, can develop a walk from the primitive pace, and can either maintain it at will or, shifting from the right to the left lateral support, show a high degree of division of functional labor in the use of the limb. In like manner, when an animal wishing to increase his rate of speed can by election leave the walk to amble, gallop, trot, or pace, an advantage is secured over another animal in which the choice is either withheld or limited.

Of the two pairs of limbs, the front one is the more adaptive for the reason that it is capable of so many secondary modifications, as grasping, flying, etc. We may assume by analogy that the fore limbs are active in changing gaits, the hind limbs acting in a more rigid and uncompromising manner. (See p. 51.)

---

* According to F. C. Selous, Proc. Zool. Soc. London, 1881, p. 726, *Rhinoceros simus* can gallop.

† Antelope and Deer of North America, p. 270.

THE MOVEMENTS OF THE HUMAN SUBJECT.

The motions of the hands and feet are essentially those of the quadruped. In the series 258, the man rising from a recumbent position on the ground first uses the left arm for support, subsequently transfers the weight to the right, and after gaining the upright position retains the hand last named as though it were in the first stage of the recover, as seen in the foot of the quadruped. The dictum made in connection with the flexor muscles aiding the foot of the quadruped to maintain itself on the ground is here exemplified (p. 42). It is interesting to note that the moment the hand leaves the ground it becomes markedly flexed.

Gait.

In series 69 the position of the arms (being thrown back at the time that the hind leg of the opposed side is also thrown back) sustains the assertion that the motion of the limbs of man is heterochiral. The chief distinction between the gait of man and that of the quadruped is the degree of torsion of the vertebral column.

For a good illustration of this torsion the figure of the boxer (series 333) may be consulted.

The Action of the Disengaged Limb.

In series 311 and 312, while the right hand is engaged in throwing a weight the left is in a state of great tension. The limb is extended and the fingers are abducted. The fingers remain extended save at the terminal phalanges, and show activity of the interossei muscles. The position just named is suggestive and presents a *raison d'être* for the following reflections:

First, that the unemployed arm is thrown into a state of excitement because of the intentive strain on groups of muscles in close co-ordination with them,—viz., the muscles of the limb of the opposed side. The limb cannot repeat the exact attitude of the one in use, and naturally tends to take on the form of the primal limb. This form is that of a motion as though it were on the ground and serving the purposes of progression. As is known from the study of the quadruped, the limb after extension passes gradually into flexion. It is evident that the interossei muscles would not flex the terminal phalanges if the foot were on the ground, but would simply hold the ends of the toes firmly to the plane on which the animal is moving. The toes are also markedly abducted in the same position. The attitude taken by the human fore limb is precisely that of the fore limb of the quadruped engaged in terrestrial progression, and is modified in its action solely by the circumstance that it "claws the air." The function of the interossei would appear to be the retention of the phalanges of the terminal row against the ground as the last act of the behavior of the foot in the first stage of the recover.

Second, that the occurrence of convulsive seizures, in which the limb assumes the position just described, may not necessarily be the result of an irritation of the centres which control the co-ordination, but an evidence of a profound irritation elsewhere in the economy. This phenomenon properly belongs to the group of influences described by Drs. Mitchell and Lewis.*

To a still greater extent than in the quadruped the superficial muscles are conspicuous. It is of interest to note that the sartorius muscle may be tense even when the foot is on the ground, as is observed in the figure of a man lifting a weight (series 322, Fig. 12). In various attitudes the several portions of the trapezius muscle are contracted. In series 323, Fig. 9, in the figure of a man engaged in lifting a dumb-bell, the upper portion of the dorsal division of the muscle is enormously enlarged.

Oscillation.

The single series designed to illustrate the movements of the pelvis in walking is not in all respects satisfactory.

Since the index at first inclines to the right side, which corresponds to the limb which is off the ground, the idea is at once suggested that the inclination is greatest towards the side not bearing weight, especially when it is noted that the index is vertical and parallel to the vertebral column when the legs are both on the ground. But when the left foot is lifted the lever remains in the vertical position, and no definite conclusion as to the effect of the movements of the limbs on the oscillation of the level is determined.

---

* Philadelphia Medical News, 1885, vol. xlviii. p. 169.

FLIGHT.

The flight of the bird is illustrated in series 755, 756, 757, pigeon; 758 to 762, cockatoo; 763, 764, hawk; 769, 770, eagle. The following conclusions have been drawn from the photographs:

The wing is extended upward from the horizontal position by the deltoid and latissimus dorsi muscles to a line which is perpendicular to the body, and is quickly again depressed to the horizontal position by the pectorals. (See p. 40.) This constitutes the first stage of the "stroke."

"Recover" is initiated by an inward rotation of the humerus, semiflexion of the wing at the elbow (the pinion remaining extended and directed obliquely downward and outward), and is carried well forward to a degree sufficient, when seen in profile, to conceal the head. In this position the primaries are semi-rotated so as to present the least amount of surface to the air in the direction in which the bird is moving. The impetus excited by the "stroke" carries the bird upward and forward. In the second stage of "recover" the humerus is rotated outward, the arm is quickly raised, the primaries restored to the position seen in the bird at rest, and the wing is a second time ready for the "stroke." (See p. 38.)

In the eagle and the hawk the legs are in the position of the "stroke" when the wings are similarly placed. During the "stroke" the legs move backward. This position continues during the "recover" of the wing, so that the time of the "recover" is also that of the "recover" of the leg.

BIOLOGICAL DEPARTMENT
UNIVERSITY OF PENNSYLVANIA,
July 1, 1887.

APPENDIX.

A LIST OF THE QUADRUPEDS AND BIRDS REPRESENTED IN THE PHOTOGRAPHS.

BY EDWIN A. KELLEY.

Series 563–657. Horse—Equus caballus, Linn.
" 658–664. Mule, hybrid—♂ E. asinus and ♀ E. caballus.
" 665–668. Ass—Equus asinus, Linn.
" 669–672. Ox—Bos taurus, Linn.
" 673–675. Sow—Sus scrofa, Linn.
" 676–679. Goat—Capra hircus, Linn.
" 680. Oryx—Oryx leucoryx, Pall.
" 681. Virginia Deer—Cariacus virginianus, Gm.
" 682. Fallow Deer—Dama vulgaris, Gray.
" 683. Virginia Deer—Cariacus virginianus, Gm.
" 684–691. Fallow Deer—Dama vulgaris, Gray.
" 692–695. Elk—Cervus canadensis, Schreb.
" 696. Eland—Oreas canna, H. Smith.
" 697–698. Dorcas Gazelle—Gazella dorcas, Linn.
" 699–700. American Bison—Bison americanus, Gm.
" 701–702. White-tailed Gnu—Connochætes gnu, Zimm.
" 703–715. Dog—Canis familiaris, Linn.
" 716–720. Cat—Felis domestica, Bris.
" 721–726. Lion ♂—Felis leo, Linn.
" 727–728. Lion ♀— " " "
" 729–730. Tiger ♀—Felis tigris, Linn.
" 731. Jaguar—Felis onca, Linn.
" 732–735. Elephant—Elephas indicus, Cuv.
" 736. Dromedary—Camelus dromedarius, Linn.
" 737–742. Bactrian Camel—Camelus bactrianus, Linn.
" 743. Guanaco—Lama huanacos, Mol.

Series 744–745. Raccoon—Procyon lotor, Linn.
" 746. Capybara—Hydrochœrus capybara, Erxl.
" 747–749. Chacma Baboon—Cynocephalus porcarius, Bodd.
" 750. Sloth—Cholœpus hoffmani, Peters.
" 751–754. Kangaroo—Macropus giganteus, Shaw.
" 755–757. Common Pigeon—Columba livia, Bris., var. domestica.
" 758–762. Cockatoo—Cacatua galerita, Lath.
" 763. Red-tailed Hawk—Buteo borealis, Gm.
" 764. Fish-Hawk—Pandion haliaëtus, Linn, var. carolinensis, Gm.
" 765–768. Turkey-Vulture—Cathartes aura, Linn.
" 769–771. American Eagle—Haliaëtus leucocephalus, Linn.
" 772–773. Ostrich—Struthio camelus, Linn.
" 774–775. Adjutant—Leptoptilus argala, Lath.
" 776–780. Groups of birds, as follows:
Muscovy Duck—Cairina moschata, Linn.
Canada Goose—Branta canadensis, Linn.
Blue Goose—Anser cœrulescens, Linn.
Bar-headed Goose—Anser indicus, Lath.
Chinese Goose—Anser cygnoides, Linn.
European (Mute) Swan—Cygnus olor, Gm.
American Swan—Cygnus americanus, Sharpl.
Black Swan—Cygnus atratus, Lath.
Sand-hill Crane—Grus canadensis, Linn.
Australian Crane—Grus australasiana, Gould.
Demoiselle Crane—Anthropoides virgo, Linn.
Crowned Crane—Balearica pavonina, Linn.
White Stork—Ciconia alba, Bechst.
Black Stork—Ciconia nigra, Linn.
Adjutant—Leptoptilus argala, Lath.
etc.
" 781. Chicken—Gallus bankiva, Temm., var. domesticus.

The following list embraces the series that are mentioned by number in the text of the report:

258	576	612	633	675	695	729	747	760
311	579	616	637	677	703	730	750	761
312	581	617	641	680	704	733	755	762
322	594	619	642	681	707	743	756	763
323	595	622	649A	682	709	744	757	764
333	601	631	658	682A	720	745	758	769
574	602	632	670	683	728	746	759	770

A STUDY

OF SOME

NORMAL AND ABNORMAL MOVEMENTS

PHOTOGRAPHED BY MUYBRIDGE.

BY

FRANCIS X. DERCUM, M.D., PH.D.,

INSTRUCTOR IN NERVOUS DISEASES, UNIVERSITY OF PENNSYLVANIA.

IN the following study the writer has made no attempt to give a systematic description of any of the plates, but only to present the salient features of the more important clinical subjects investigated. Naturally these subjects are of greater interest to neurologists than to others. The normal walk, however, possesses an interest that is general, and the writer believes that the Muybridge method has in this field, as in others, yielded new results.

In addition to the serial photographs which have been published, Mr. Muybridge kindly made, upon request, a number of clinical photographs by means of a camera armed with a fenestrated wheel,—*i.e.*, a Marey's wheel. Some of these pictures are here reproduced.

Thanks are due to Drs. S. Weir Mitchell, William Pepper, H. C. Wood, Charles K. Mills, and James H. Lloyd for the opportunity of photographing various patients under their care.

The writer desires also to express his obligations to Mr. Muybridge for his great and many courtesies, which necessitated not only encroachment upon his most valuable time, but also the occasional adoption of methods not included in his original enterprise.

Regarding the various trajectories depicted in this monograph, it is of course not claimed that they are *absolutely* accurate, but merely as close an approximation to the truth as the method of research permits.

The Normal Walk.

In order that the various abnormal gaits should be properly understood, a study of the normal walk had first to be made. In this connection it is hardly necessary to refer to the work of the Weber brothers nor to the earlier results of Marey and Carlet, for they are already the common property of schools and text-books. In more recent years, as is well known, Marey invoked the aid of photography to enable him to confirm and add to his original results obtained by the graphic method. His photographs, however, were confined simply to the analysis of the vertical and forward movements of various points of the body, and they gave no information whatever of the direction or extent of the lateral sway. Marey was well aware of this, and by a most ingenious application of the stereoscope to his photographic wheel he endeavored to remedy this defect. The pictures that he obtained are exceedingly interesting, and when examined stereoptically give one the impression of an undulating white band extending through space, the undulations being in three directions, forward, vertically, and laterally. His achievement was indeed a brilliant one, and yet the pictures do not admit of a detailed study of the curves.

However, the method invented by Muybridge, of making simultaneous serial photographs of a moving man or animal from two points of view at right angles with one another, has yielded pictures furnishing all of the elements necessary to determine the various paths of motion. It must, however, be admitted that while by this method the lateral sway is quite definitely ascertained, there is a slight loss in the accuracy of the curve of the vertical and forward movements, and this arises from several causes. In the first place, a possible and probable source of error is slight irregularity in the intervals of time between the successive photographs of a series. No one would pretend that the same accuracy as regards regularity in the sequence of exposures could obtain in a serial battery of cameras as in such an apparatus as used by Marey, in which the sequence of exposures was determined by the fenestra of a revolving wheel. Notwithstanding, repeated chronographic measurements made by Muybridge showed the irregularities of these intervals to be exceedingly small even for very rapid movements. Therefore in a movement relatively so slow as the walk they can be practically discarded.

Another cause for slight loss of accuracy arises from the fact that the person walking cannot be in exact apposition with the lateral background which is used as a scale. This error is of course lessened by the use of lenses of long focus,—*i.e*, in the lateral battery of cameras, which was placed upwards of fifty feet from the track. Being by this means so much diminished, it can also be discarded. A third source of inaccuracy is due to the fact that the various phases of any movement photographed cannot be, except by improbable accident, directly in front of the centre of each corresponding camera. Therefore each individual phase, instead of being photographed from a position exactly at right angles to the background, is doubtless in the majority of instances photographed at an angle varying slightly from the right angle. Furthermore, for the various phases, this angle is a variable and indeterminate quantity. However, it must again be insisted that for the majority of movements, especially slow movements, such as walking, this error is also small, though it is obviously greater than the second error just mentioned.

Taken all in all, while the serial method gives slightly less accurate results regarding the rise and fall and onward movement of a limb, it more than compensates for this loss—which in itself is slight—by enabling us to determine the amount and direction of the lateral sway. Furthermore, apart from the mere determination of trajectories, two photographs of any one phase of movement taken at right angles give us an opportunity for the study of the action of a part which a wheel-photograph can in no way furnish.

Having considered the various sources of error, let us take up the study of the walk in Plate 1. The model was a young man of medium height and was photographed while taking a long step.

By means of a transparent scale identical with that of the background, the vertical and forward movements of various points of the body were readily studied. The lateral sway was determined by means of a transparent scale based upon the broad divisions of the background at the end of the track,—that is, the distance between the heavy white lines, which is equivalent to thirty centimetres, was taken as the basis of the scale; but for greater accuracy this distance was divided into ten parts, instead of into six, as in the original background. A plan of the track was then made (see Diagram 1) and the scale corrected for the error in perspective for the area in which the step was taken. In Diagram 2,

FIG. 1.

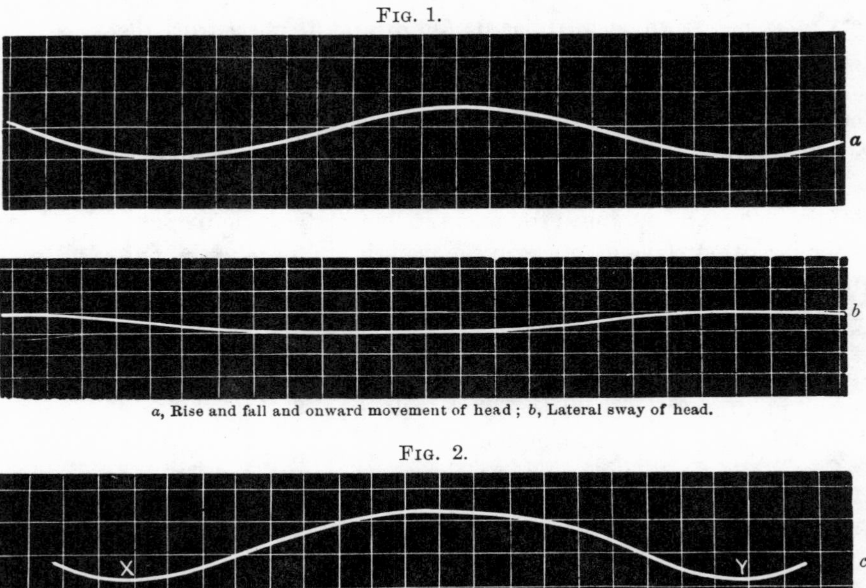

a, Rise and fall and onward movement of head; *b*, Lateral sway of head.

FIG. 2.

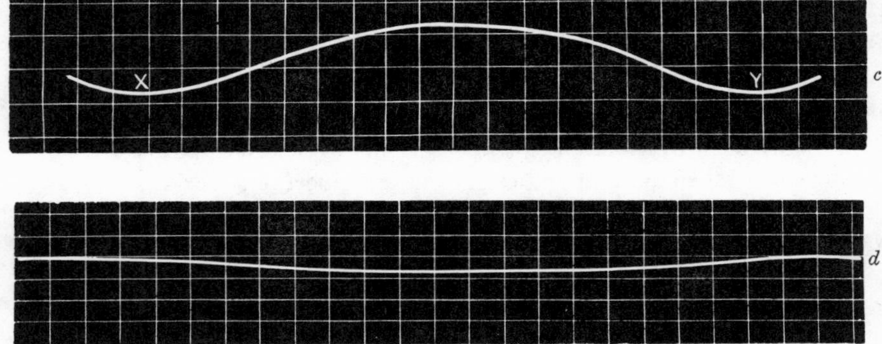

c, Rise and fall and forward movement of right hip; *d*, Lateral sway of hip.

the actual size of the track and the normal scale are in red lines, while the corrected scale is in black lines.*

It should have been stated that the study of the lateral sway was made upon the middle series of Plate 1, as here the images and background are comparatively sharp.

---

* See explanation of Diagrams.

DIAGRAM 1.

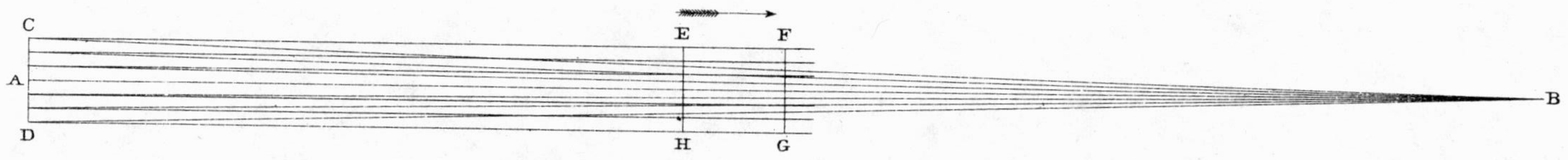

E

A

H

EXPLANATION OF DIAGRAMS.

In the following explanation the words "lateral scale" are used to designate the divisions of the background opposite the lateral battery of cameras. The words "terminal scale" are used to designate the divisions of the background at the end of the track.

It should also be noted that in making the study of the lateral sway, the transparent scale was for convenience so applied that one of its heavy lines coincided with the central line of the terminal scale. This line is in the centre of the broad division occupying the middle of the background. The line AB of Diagram 1 and 2 is, therefore, a projection of this line along the course of the track, and the divisions and subdivisions of the scale are projected in relation with this central line.

DIAGRAM 1.

The line AB equals the length of the track. The divisions of the line CD equal the divisions of the terminal background, equivalent to thirty centimetres each. B equals the position of the upright battery of cameras by means of which the middle series of pictures of Plate 1 of Muybridge's photographs were made. EFGH equals the portion of the track on which the movement photographed was made. The arrow indicates the direction in which the model moved.

DIAGRAM 2.

The portion of the track marked EFGH in Diagram 1 is in Diagram 2 represented much enlarged. The horizontal red lines are noticed to be of two kinds, heavy and light. Of these the heavy lines represent the broad divisions of the terminal scale projected along the track. They are thirty centimetres apart. In the original scale this space is subdivided into six others of five centimetres each. In the diagram, however, for greater accuracy it is subdivided into ten parts of three centimetres each, as also in the transparent scale used in making the actual study on the photographs.

The vertical lines in the diagram represent the vertical divisions of the lateral scale projected across the track. They are, therefore, five centimetres apart.

The black and slightly converging lines constitute the corrected scale upon which the measurements made on the photographs are registered and on which the curves are constructed. The method by means of which this scale is determined is illustrated in Diagram 1.

The heavy black line marked f is intended to show the extent and direction of the lateral sway of the left foot. The lines b and d are intended to show the same elements for definite points on the head and hip respectively.

It is evident that the direction of the step was not exactly the same as that of the track, but that it was slightly oblique.

In Fig. 1, the line *a* represents the rise and fall and onward movement of the white button in the cap worn by the subject, while the line *b* represents the amount and direction of its lateral sway. As in the trajectory of the pubis determined by Marey, the number of vertical oscillations appear to be double those which take place in a horizontal direction.

In Fig. 2, the rise and fall of the right superior spinous process of the ilium is represented in the line *c*. It is similar in general course to the line *a*, but the amplitude of the wave is much greater. This increase in amplitude is doubtless in part due to the meagre lateral oscillation of the ilium during the time the right foot is fixed upon the ground. In fact, in the line *d* the amount of curve is exceedingly small,—*i.e.*, during the time the ilium is describing the arc *xy* (line *c*).

In Fig. 3, the line *e* represents the rise and fall and forward movement of the left foot, the internal malleolus being taken as a

FIG. 3.

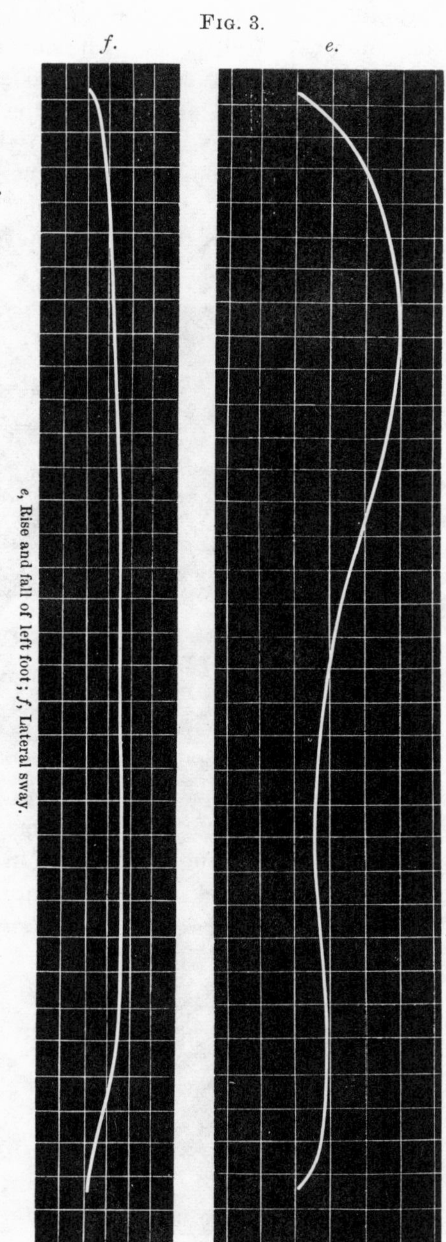

e, Rise and fall of left foot; *f*, Lateral sway.

definite point. It is seen at a glance that this curve is made up of a number of elements. The curve from Y to S is formed by the maleolus sweeping upward on an arc the radius of which centres in the ball of the great toe. At the next instant the centre of rotation is transferred to the *tip* of the great toe, and a change in direction is noticed, therefore, at the point S. The balance of the curve up to the point Z is the result of the flexion of the leg and of the forward movement of the pelvis; thence

FIG. 4.

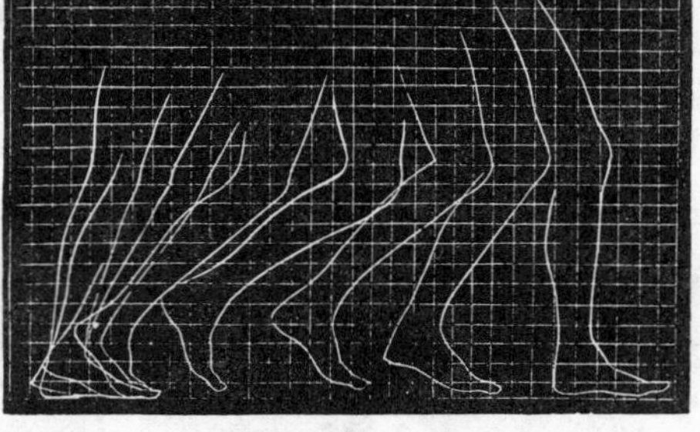

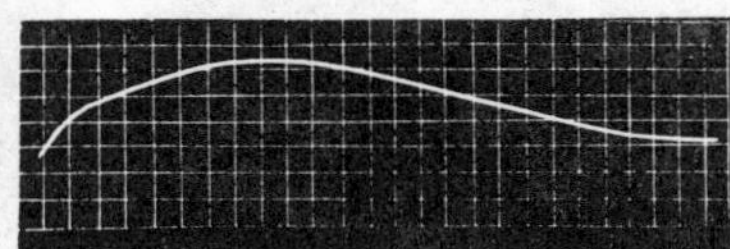

Tracing and trajectory constructed from Plate 8. The trajectory is supposed to pass through the external malleolus. (Incomplete.)

to the final impact of the heel upon the ground the curve is the resultant of a complex movement, in which three principal elements are distinguishable: first, a *pendulum movement*, second, a *fall*, and third, a *forward movement*, the latter being due to the movement forward of the body as a whole. The first two elements are those of a cycloid, and the foot therefore falls to the ground, other things equal, *along the line of swiftest descent.*

FIG. 5.

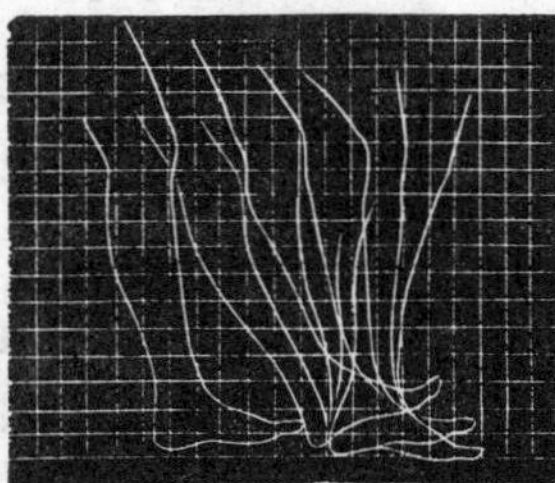

Left foot, Plate 8.

A fourth element is observed in the slight secondary rise occurring in the curve just previous to its termination. The significance of this rise is as follows. We notice that the heel of the passive leg in swinging forwards in its cycloid-like descent does not immediately strike the ground, but that just previous to the impact it again makes a slight ascent. This is shown not only in the curve, but also in all of the plates illustrating the normal walk. (See, also, Fig. 5.) We should observe at the same time that the rate of motion is not by any means uniform. It is relatively slow in the beginning, but steadily increases in velocity until the step is almost completed, when a slowing of movement again takes place. This slowing of movement is coincident with the secondary rise. The impact of the heel upon the ground is then made without any waste of force and with a minimum amount of jar. When in addition we reflect that the heel is but a portion of a moving lever and that the muscles attached to this lever are elastic, we realize that the jar of impact is indeed reduced to a minimum.

The curves of the lateral sway are also exceedingly interesting. The fact that the lateral sway of the head is greater than that of the hip is exactly in keeping with what we should expect on *a priori* grounds. The pelvis is relatively fixed and not nearly so free to move as the head or upper portions of the trunk, and, secondly, the tilting of a column is naturally exaggerated at its upper and free end.

The lateral sway of the foot again presents an interesting interpretation. Its extent is less than most of us would have expected to find, being in fact very small. This signifies undoubtedly that the extent of the lateral sway is, other things equal, always the least possible consistent with the working of the limb; because, first, the greater the lateral sway the greater the loss of time consumed in again bringing the foot to the ground; and, secondly, the greater the lateral sway of the foot the greater the lateral sway of the body as a whole, and the greater the loss of time and force at the expense of the forward movement.

Finally, taking all of the trajectories together, it is evident that the movements in the three directions, forward, vertically, and laterally, are correlated. Secondly, the prime object of the walk being movement in a forward direction, it follows, other things equal, that the greatest economy of time and force obtains when the vertical and lateral movements are reduced to a minimum. This is apparently the case in the normal walk.

The Gait in Locomotor Ataxia.

This gait was studied in Plates 546, 549, 550, 554, and 560. All of the cases photographed were typical of locomotor ataxia. The patient upon whom the trajectories were studied is the one of Plate 560. His history is briefly as follows:

H. R., aged thirty-four, a clerk by occupation, first noticed that his general health was failing in the fall of 1883. He was at first treated for general neurasthenia, but soon developed sciatic pains and gastric crises. These last were very severe. He then disappeared from observation for two years. When he again returned he was markedly ataxic, both in the arms and legs. He had parasthesia of the soles of the feet, contracted pupils, loss of the patellar reflex, retardation of sensation, etc.

In Fig. 6 is represented the right foot from the time of leaving to again striking the ground. It needs but a superficial glance to show that it differs markedly from the normal foot in

FIG. 6.

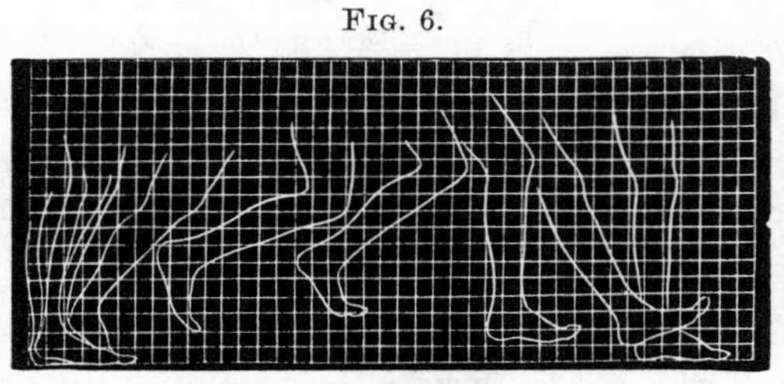

Locomotor ataxia, right foot.

its course. Compare, for instance, Figs. 4 and 5. It differs more widely still from the foot seen in Fig. 9, that of lateral sclerosis.

The direction and extent of the lateral sway was determined by means of a transparent scale corrected for perspective for each position of the foot. (See Fig. 7.) The external malleolus is selected as a relatively fixed point. The curve evolved is represented in line *a*, Fig. 8, whilst the rise and fall and onward movement are shown in line *b*.

The striking difference between line *a* and the corresponding curve of the normal gait consists in its irregular and extensive outward sway, while in the line *b* the difference consists in the increased amplitude of the curve and in the irregularity of the downward fall. Line *b* is made up of the same elements as the

FIG. 7.

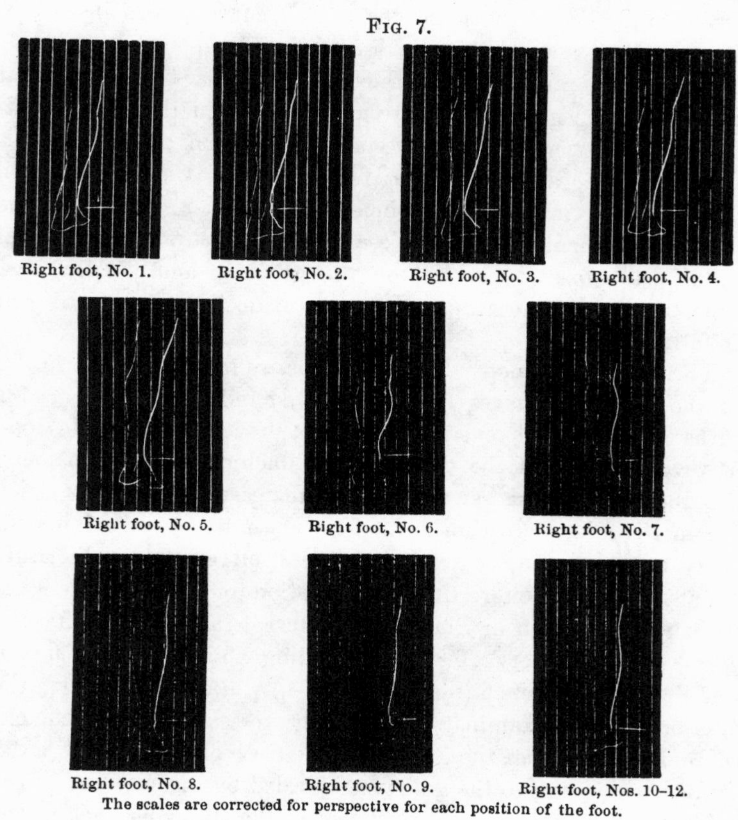

Right foot, No. 1. Right foot, No. 2. Right foot, No. 3. Right foot, No. 4.

Right foot, No. 5. Right foot, No. 6. Right foot, No. 7.

Right foot, No. 8. Right foot, No. 9. Right foot, Nos. 10–12.

The scales are corrected for perspective for each position of the foot.

corresponding line of the normal gait; namely, first, a short curve depending upon the upward rotation of the malleolus upon a

FIG. 8.

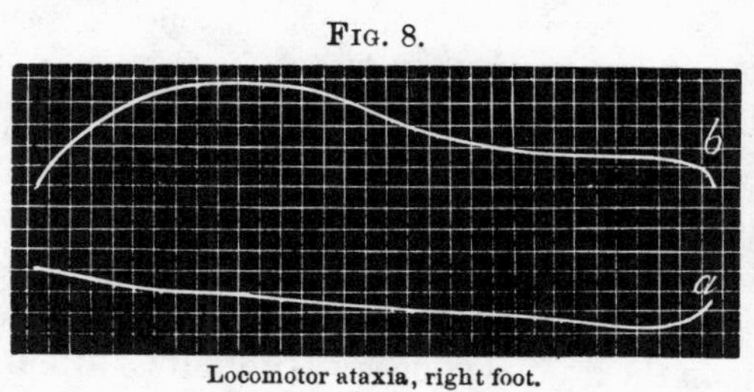

Locomotor ataxia, right foot.

The trajectories are supposed to pass through the external malleolus.

radius centring in the ball of the great toe; secondly, a more ascending curve depending upon the centre of rotation being transferred to the tip of the great toe. This curve passes insensibly into another, caused by the flexion of the leg upon the thigh and by the onward movement of the hip. It is to be noticed that flexion of the leg is much more marked than in the normal condition; consequently, the corresponding portion of the curve is higher. The fall of the foot to the ground we noticed, both in line *b* and in Fig. 6, is somewhat irregular; also, that the heel fails to make the slight ascent which occurs in the normal walk just before the impact on the ground; and, finally, that when the impact does take place it is made by a sudden descent or fall.

The writer does not by any means desire to impress the reader as though these curves (*i.e.*, of Fig. 8) indicate the invariable paths through which every ataxic must move his feet. He does, however, insist that the two principal factors indicated, namely, the increased lateral sway and the increased height to which the foot is raised, are constant. To these must be added irregularity of movement, as irregularity is itself an essential feature of ataxia. An interesting confirmation of the constancy of these factors is seen in Plate 550. Here, a Dane, aged fifty-two years, with a history of a long-standing and increasing ataxia, was first photographed (see upper series) while walking with the eyes open. An examination of the figures shows that although the gait differs from the normal in the two principal factors described, yet, owing to the guidance afforded by his eyes and to the evident effort he is making, as shown in the throwing back of the shoulders and the fixation of the trunk, his walk is tolerably good. In the lower series, however, he was photographed *while his eyes were closed.* A remarkable change is at once observed, and it needs no elaborate study of the trajectories to show what the chief elements of the change are. It is seen at once that the foot is raised much higher from the ground, and that the lateral sway is enormously increased; also, that the successive steps are more irregular. In this series, also, the abrupt descent of the foot upon the ground is again noticed, the entire sole impinging or flopping, as it were, upon the ground at once. In Plates 546 and 554, which represent cases of ataxia more or less advanced, the essential points of the ataxic gait are pronounced, as they are also in Plate 549.

To show how difficult it is to observe a moving limb, even when the movement is slow, it need only be stated that medical writers almost without exception describe this gait erroneously. Almost all lay stress upon rigidity of the leg and insufficient action of the knee-joint. It needs but a hasty examination of the photographs to show how utterly wrong this view is. Every one of the plates reveals the action of the knee-joint, and in fact of all of the joints, to be far in excess of the normal; and, further, the rigidity is *simulated* and not real. Doubtless this erroneous view has its origin in the difficulty of separating mentally the active from the passive phase of the leg. Plate 550, which is one of the most instructive in the collection, admirably suits the purpose of an explanation. It is seen in studying the upper lateral series that the passive leg, *i.e.*, the one suspended in the air, presents exaggerated knee-joint action and that it at no time gives even the semblance of rigidity. The moment, however, that it impinges on the ground, *i.e.*, the moments it becomes active, all action at the knee ceases. More than this, as the weight of the body is brought upon it, the extension becomes absolute, and finally the knee is, so to speak, *locked;* that is, recurved. (This is well seen in No. 6 of the series.) By means of this expedient the strain is thrown upon the bones and ligaments of the joint rather than trusted to the treacherous muscles. The rigidity is, therefore, not muscular, which, in fact, all of our knowledge of ataxia forbids. How much voluntary effort on the part of the muscles assists in steadying the limb, of course depends upon the degree to which the disease has progressed.

Regarding the action of the foot and the manner of its impact on the ground various accounts have been given, some writers maintaining that the heel strikes the ground first and considerably in advance of the plantar surface, and others maintaining that the entire sole strikes the ground at once. An examination of Plate 550 shows that the manner of impact probably depends upon the degree of the ataxia. In the upper series, in which the patient was photographed while having the guidance afforded by his eyes, the heel impinges distinctly before the sole, while in the lower series, in which the patient walked with closed eyes, the entire plantar surface impinged at the same time. Occasionally the toes came down first and sometimes even the heels, so that in advanced ataxia any part of the foot may strike the ground first.

Regarding the lateral sway of the trunk, it needs no detailed study to tell us that it is much exaggerated. In ataxics who are still able to walk comparatively well this exaggerated sway is always towards the side opposite the advancing or passive leg. However, if the ataxia be increased, as, for instance, in the lower series of Plate 550, the sway may be towards the same side, and then the patient is in danger of falling.

The Gait in Lateral Sclerosis.

This gait was studied in Plate 548. The subject was a patient of the University Hospital and was under the care of Dr. William Pepper. His history was as follows:*

"H. S., aged twenty-eight years, single, is an engineer by occupation, and had been working on a railroad in Wyoming Territory until three or four months ago, when he came to this city. When a child he had diphtheria, which was followed by dropsy and paralysis. Later he was thrown on a hot stove and severely burned, the scar being visible on the epigastrium. When twenty-two years of age he had typhoid fever. He denies having had gonorrhœa or syphilis, although intercourse with loose women is acknowledged. Careful investigation fails to reveal any trace of specific infection. He has used alcohol in considerable quantities, but has never been intoxicated. He uses tobacco in great excess, smoking as many as fifteen cigars on some days, and chewing a large plug of tobacco every twenty-four hours.

"His occupation has been an exposing one, subjecting him to extreme alternations of heat and cold. He, however, continued in good health until three years ago. In December, 1882, he first noticed a gradual loss of power in the left ankle. He states that five months prior to this he had injured the left ankle and knee by falling from his locomotive. This kept him in bed for seven days, then apparently he became entirely well. The feeling of weakness on the left side compelled him to throw most of his weight on the right leg. The trouble gradually grew worse, and in November, 1883, he noticed that occasionally the affected leg

---

* See Philadelphia *Medical Times*, October 31, 1885. Clinical Lecture by Professor Pepper.

would suddenly fly backward when he attempted to move it, and that the muscles of the left side would tremble violently in quick clonic spasm. Unless he moved very carefully this occurred at every step. He next observed that he was losing flesh. His height is six feet one and one-half inches, and his best weight has been one hundred and seventy-nine pounds. Three months ago, when this note was made, he weighed one hundred and sixty-five pounds.

"In July, 1884, the affected ankle was strained, and this was followed by swelling of the foot and leg.

"About this time, or a little earlier, he noticed that the fingers of the left hand were disposed to contract, so that the hand would remain closed until the fingers were forcibly extended. The contraction was increased by fatigue. This gradually increased until it involved the hand, forearm, arm, and shoulder."

At the time at which the photograph was taken, the summer of 1885, the patient presented in addition the following symptoms: He habitually carried the left arm (*vide* Plate 548) in the semi-flexed position assumed by patients having lateral sclerosis or secondary degeneration of the lateral columns. The left leg was decidedly stiff. When sitting on a chair it was extended. The knee-jerk was markedly exaggerated. The gait was "spastic." The right leg also was somewhat affected. It was somewhat resistant to flexion and the knee-jerk was evidently increased.

FIG. 9.

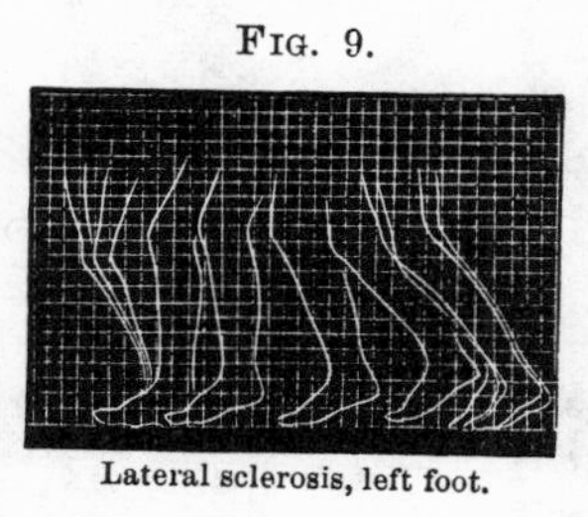

Lateral sclerosis, left foot.

FIG. 10.

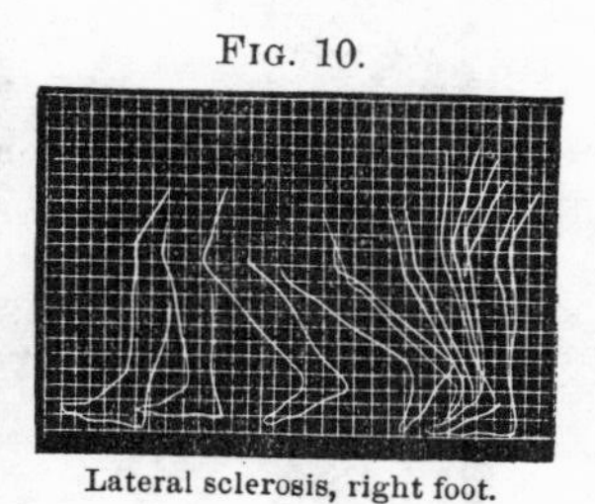

Lateral sclerosis, right foot.

The gait was studied in detail in the lower series of Plate 548. Fig. 9, as well as the lateral series in the plate, shows, in the first place, that the left foot scarcely left the ground, that the heel only was elevated, and that the ball of the foot and the toes were merely raised sufficiently to permit them to slip or scrape along the surface; also that the leg as a whole was very little flexed.

FIG. 11.

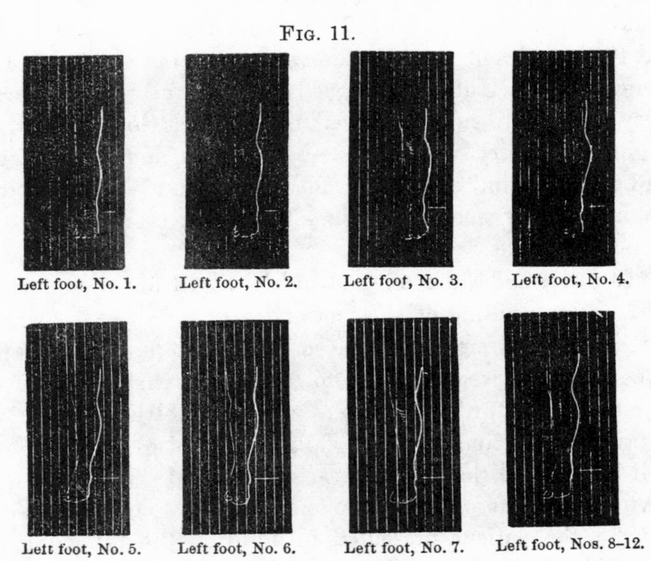

LEFT FOOT.

The scale is in each instance corrected for perspective. From 7 to 12 there is absolutely no lateral sway, as the foot is fixed firmly on the ground. The curved line represents the external outline of the leg and foot. The point A, No. 1, shows the position of the malleolus.

FIG. 12.

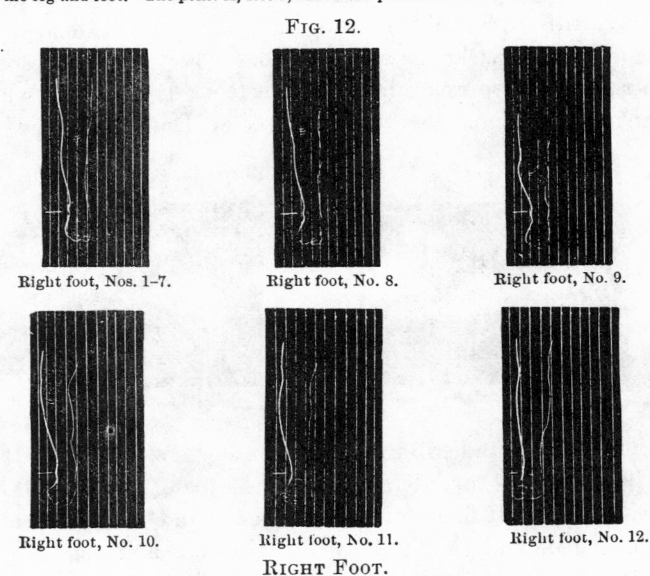

RIGHT FOOT.

As in Fig. 10, each scale is corrected for perspective. From 1 to 7 there is no variation, and from 8 to 12 the variation is comparatively small. The external malleolus was as before the point on which the study of the trajectory was based.

In the other leg the spastic condition, though present, was not so much marked. (See Fig. 10.) It is here noticed that the foot is raised clearly off the ground, and that the leg is flexed, and yet neither of these factors is as marked as we find it in the normal condition.

The trajectories were determined, and are represented in Figs. 13 and 14.

The amount and direction of the lateral sway of each foot was determined by the use of a transparent scale corrected for perspective for each position of the foot. The external malleolus was the point used in making each measurement. (See Fig. 11, Nos. 1 to 12, and Fig. 12, Nos. 1 to 12.)

FIG. 13.

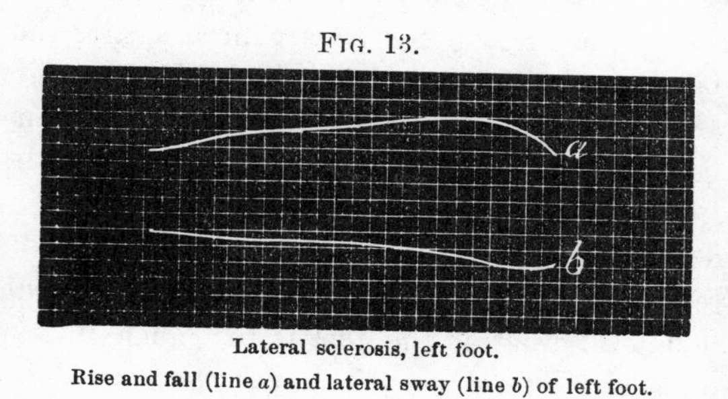

Lateral sclerosis, left foot.

Rise and fall (line *a*) and lateral sway (line *b*) of left foot.

In Fig. 13 the upper line, *a*, shows the rise and fall and onward movement of the foot, whilst the lower line, *b*, shows the amount and direction of the lateral sway.

FIG. 14.

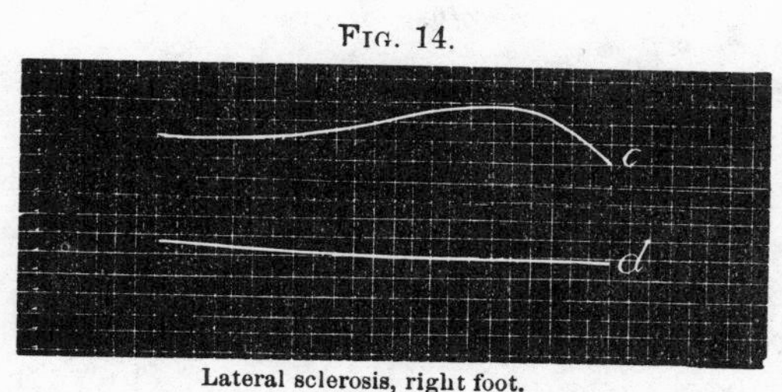

Lateral sclerosis, right foot.

Rise and fall (line *c*) and lateral sway (line *d*) of right foot.

In Fig. 14 the lines *c* and *d* represent the same factors of the right foot. It is noticed in the left foot that the rise and fall as well as the lateral sway are very small. It will be remembered that the left leg was the one most diseased. It was typically affected. In Fig. 14 the rise and fall are more marked, and this again is in harmony with the fact that the right leg was much less affected than the left. On comparison, however, with the rise and fall of the normal leg (see Fig. 5), these factors are seen to be much less pronounced.

Again, the left foot at the completion of its movement is brought too far within the median line, and thus gets in the way of the right foot. The latter in order to prevent a fall is brought too far outside of the median line, and thus the direction of the walk as a whole is changed.

As regards the lateral sway of the trunk, it is of course grossly exaggerated, and always takes place towards the side opposite the advancing leg. It is, in fact, only by means of this grossly exaggerated sway that the leg can be advanced at all; the sway tilts the pelvis and thus assists in raising the limb from the ground.

As regards the relative extent of the lateral movement of the trunk, it is always greater towards the side of the less affected limb.

The gait in lateral sclerosis is so very slow that medical writers have in general given accurate descriptions of it, though the extent of the lateral sway of the foot is too much insisted on. In all probability the lateral sway is never very marked when both legs are affected or affected about equally. It, however, becomes much exaggerated, as we shall see below, when the disease is confined to the lateral column of one side.

Various other Spastic Gaits.

In Plate 553, are presented the photographs of an interesting case, the history of which is briefly as follows: C. M., aged twenty-four, a laborer, sixteen months ago fell into a pit eight feet deep, striking on the top of his head; he was unconscious for four hours, and experienced upon awakening a severe pain in the back of the neck which increased upon motion. He had absolute paralysis of both legs for six weeks, and of the bladder and rectum for four weeks. The arms, especially in the act of flexion, were markedly paretic. He did not regain the power of walking for nine weeks. He had improved to a certain degree, but failed to progress. At the time of being photographed the pupils were slightly unequal, the right being the larger. No affection of special senses except impairment of hearing in right ear was noted. No anæsthesia existed anywhere. The *legs* were *paretic* and spastic. No specific history could be elicited.

The spastic condition of the legs is well seen in the photographs, though the gait is by no means a typical spastic gait, such as seen in Plate 548 (lateral sclerosis). The feet are raised but little from the ground, and appear to slide along the surface. The lateral sway, though, is greater than in typical disease of the lateral columns. In the latter, the meagre outward movement of the foot is brought about by a forced and extreme swaying of the trunk, and not by the action of the leg itself.

In Fig. 543 is found another illustration of an atypical spastic gait. The subject was an hysterical girl.

It is seen by even a cursory examination that the feet, though not raised to the normal height above the ground, are yet raised to a height greater than that seen in typical lateral sclerosis. The amount of the lateral sway, too, is also relatively large. There is some tendency to scrape the feet along the ground, especially the right one, but there is a decided outward movement.

Other instances of spastic gait are seen in Plates 547 and 552. Both subjects were old hemiplegics with marked secondary contracture. In both instances the paralyzed leg is quite stiff, little or no flexion taking place at the knee. It is also noticed that the foot is here, almost or quite, raised from the ground by the enormous swaying of the trunk towards the sound side, to which additional support is given to receive the sway by means of the crutch. It is significant, therefore, that the crutch be carried on the sound side. In these figures is also seen, precisely as in typical laterai sclerosis, the exaggeration of the normal tendency of bringing the outer edge of the foot to the ground in advance of the sole. This is especially shown in the lower series of Plate 552.

These instances of what might be called unilateral sclerosis differ, therefore, from double lateral sclerosis chiefly in the great exaggeration of the lateral sway.

In Plate 542 still another example of spastic gait is found, but it is not typical. The history of the case in brief is as follows: J. C., female, aged forty-one, commenced to have pain in the back some fourteen years ago. The pain has steadily continued. Soon after its commencement she noticed some difficulty in walking. The legs "seemed to get stiff and to tremble." Afterwards jerking of the hands, arms, shoulders, and head commenced. This jerking, or chorea, is still present, though less marked than before. She has, however, a choreic or spasmodic affection of the muscles of respiration and phonation, as her breathing and her speech are very irregular. Her speech is not infrequently interrupted by a sudden gush of inspiration, which produces an ascending note. She has no chorea of the legs, but her gait is markedly spastic, and she walks with great difficulty. Her face, too, is somewhat choreic, and her tongue deviates slightly to the left.

The case was doubtless one of disseminated sclerosis in which the lateral columns were markedly affected.

An examination of the plate shows that she barely, if at all, raises the feet from the ground. The amount of lateral sway, too, is small, and her steps are exceedingly short.

In Plate 541 is seen the walk of an extremely choreic girl,—a chorea which had lasted from early infancy and appeared to be of organic origin, probably a disseminated sclerosis. It is therefore allied somewhat to the preceding case. At first the gait appears spastic. However, the leg which is at one instant rigid and extended is at the next instant flexed. It is interesting to note the spastic condition of the right arm, which, during the step, is rigidly extended along the side of the body. The head and neck and even the toes are violently contorted. Of course in such a case no two steps could be alike, and the gait could only be termed irregular.

Other Abnormal Gaits.

In Plate 551 is represented an epileptic suffering from a spastic hemiplegia which is more pronounced in the arm than in the leg, and which dates from early childhood. The gait is in some respects a spastic gait.

Somewhat similar to it is the gait, seen in Plate 561, of a rachitic and hydrocephalic subject. Here both legs are decidedly spastic.

In Plate 559 we have an instance of partial paraplegia. The subject was a druggist in a country town, who had, several years before, been thrown from a horse and had received a severe contusion of the back. He gradually lost power in his legs until almost unable to walk. Syphilis, though denied, was suspected as a concomitant factor, and the man made marked improvement under large doses of potassium iodide.

It is to be noticed that the feet are thrown forward in a passive or pendulum-like manner, and that the weight is not trusted to a leg until it has been *locked,—i.e.*, until the knee has been thrown far backward (incurved),—so that collapse from sudden flexion becomes impossible.

In Plate 555 is seen the gait of a case of monoplegia. The monoplegia is due to a commencing muscular atrophy affecting the left thigh and leg. It is seen that the diseased leg is thrown forward (see lower series) in the same helpless or pendulum-like manner as in the case of paraplegia; also (see upper series) that the weight of the body is not thrown upon it until the knee has been well *locked.* This is in marked contrast with the action of the normal limb. (See plates of normal walk.)

Again, the strange gait of Plate 558 is worthy of a word. It is that of a case of stuporous melancholia. His history is as follows. He is twenty-five years old. For thirteen years he worked as a type-finisher, being constantly exposed to fine particles of dust mixed with lead. Three years ago he had an attack of acute lead-poisoning, with marked wrist-drop. A year later he had a period of excitement, with evidences of insanity. He then had hallucinations of sight and hearing, and had delusions of persecution; thought his sister was trying to poison him, and that his fellow-workmen were constantly endeavoring to have him discharged. After being in this condition for about six weeks, he became sullen and stuporous; he would make no effort for himself; had to be fed and had to have all of his wants attended to by his friends. His stuporous condition largely persists. He does not speak and is apt to remain in positions in which he is placed. His gait might be described as a crouching shuffle. It is certainly in keeping with his mental condition.

Another remarkable gait is that of a case of infantile paralysis, which is depicted in Plate 539, and in which the child has adopted the method of walking on all fours. The foot-falls, so to speak, occur in the same order of succession as they do in the walk of a quadruped. The case was under the care of Dr. James H. Lloyd, of the University.

Artificially-Induced Convulsions. Plates 544 and 545.

In order that these plates may be understood, it will be necessary to quote from the paper* in which these convulsions were first described.

"Our experiments were performed by subjecting a group of muscles to a constant and precise effort, the attention being at the same time concentrated upon some train of thought. The position we most frequently adopted was the following: The subject being seated, the tips of the fingers of one or both hands were so placed upon the surface of a table as to give merely a delicate sense of contact,—*i.e.*, the fingers were *not allowed to rest* upon the table, but were maintained by a constant muscular effort *barely in contact* with it. Any other position involving a like effort of constant muscular adjustment was found to be equally efficient. Any one object in the room was now selected and the mind fixed upon it, or some subject of thought was taken up and unswervingly followed.

"After the lapse of a variable period of time, extending from a few minutes to an hour, . . . tremors commenced in the hands. These tremors became rapidly magnified into rapid movements of great extent, sometimes to and fro, sometimes irregular. If the experiment was now continued, the muscles of the arms, shoulders, back, buttock, and legs became successively affected, and the subject was frequently thrown violently to the ground in a strong general convulsion.

* * * * * * * * * *

"Seizures equalling in violence a general convulsion were by no means induced in all subjects, and were generally the result of experiments repeated many times during the same evening."

The subject employed in the experiments from which the photographs were made was a professional artist's model, a woman, aged thirty-five, of indifferent or phlegmatic temperament. The conditions of the experiment were much less favorable, of course, than they would have been in a private room. They were performed in the large open photographic yard and amid the distracting circumstances of strange and unusual preparation. How-

---

* "On the Artificial Induction of Convulsive Seizures," by F. X. Dercum, M.D., and A. J. Parker, M.D.: Journal of Nervous and Mental Disease, vol. xi., No. 4, October, 1884

ever, as both the plates and figures show, the results were by no means unsatisfactory.

In Fig. 15 the position of the subject and the early stages of an induced convulsion are shown. The figure is the result of a number of tracings of the serial photographs of Plate 545, lower series, being superimposed. The tremor is just being magnified into to-and-fro movements of the hands and feet. A photograph of this stage was also made by means of a "Marey wheel," and is shown in Fig. 16. The blurring of the hands and feet here show the extent of the convulsion.

FIG. 15.

Fig. 17 is the result of superimposed tracings of the upper series of pictures in Plate 545. In this instance the subject was simply seated on a chair facing the cameras. There was no table used. The model had already induced a number of convulsions, and was now asked to induce a convulsion by keeping the fingers in delicate contact with the thighs. In a short time, as shown by the figure, a most violent convulsion was the result.

FIG. 16.

The individual pictures of this series are especially interesting. The entire pose or attitude assumed by the subject is, so to speak, *hysteroidal*, while the purposeless movements of the limbs suggest those of chorea, than which, however, they are infinitely more rapid.

In Fig. 18, which is the result of a superimposed tracing of the first series of Plate 544, the subject was lying upon a mattress, and the convulsion was induced by simply attempting to keep the hands in delicate contact with the body. The convulsion, as shown by the figure, was of considerable violence.

Fig. 17.

It might have been possible by prolonging the experiment to produce still more startling results, but the unfavorable surroundings, the temperament of the subject, and the fact of her being much exhausted, forbade, as being both impracticable and unjustifiable, any further attempt.

The reader may appreciate more fully the violence of these seizures when told that of the various figures here given, each represents only a brief portion of a convulsion,

Fig. 18.

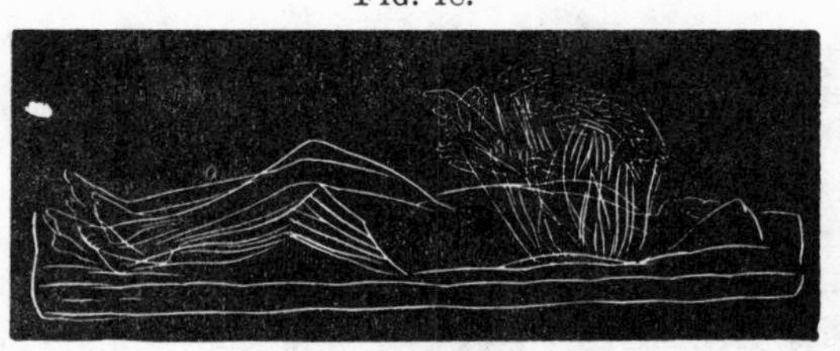

and that portion embraced within so small an interval of time as 1.8 seconds.*

Case of Functional Spasm. (*Local Chorea.*) Plates 556 and 557.

This remarkable case, some of the phenomena of which resemble those seen in the artificially induced convulsions, was first studied by Dr. S. Weir Mitchell, and reported by him, in the *American Journal of the Medical Sciences* for October, 1876, in a paper entitled "On Functional Spasms." From it we abstract the following history:

* For further particulars concerning these interesting phenomena the reader is referred to the original paper.

"R. C., aged thirty-three, watchcase-maker, married. No syphilis. Family history bad. A sister, aunt, and grandmother had palsies in middle life; an uncle had epilepsy, and a cousin dementia. He was always nervous and excitable, used no spirits, smoked moderately, and was always well until the summer of 1865, when he was two days unconscious from sunstroke, and has ever since been made weak by the heat of summer. About two years ago, in the summer of 1874, he observed that while walking the legs felt weak, and he was obliged to stop, as if to regain power; after which the right leg would drag for a time. At the same time he began to have pain in the back of the neck and lumbar spine. These pains continue. In January, 1875, he noticed a slight but increasing tremor in the left arm, and in two or three months the strange condition which I shall try to describe.

"When I first saw Mr. C. he was a healthy-looking man, of unusual intelligence, and marvellously patient under his great suffering. In sleep there was no movement; when he awakened he was conscious of the left hand being rigidly closed. In a few moments it began to twitch, the fingers moving as do those of a violin-player. The slightest movement of any other limb, speaking, or eating—even if he be fed—causes the left arm to execute a constant motion of striking the bed or his side, the limb being the while extended. When he arose and walked, this action became more violent, and so much resembled the steady, rapid movement of a pendulum, that I spoke of it at once to my assistants as a case of what might be called *pendulum spasms*. In fact, its rhythmic regularity was astonishing. Dr. Sinkler timed it, on one occasion, as one hundred and fifty-seven (in a minute); and on several others I found it always exactly one hundred and sixty. It was as accurate as the heart in its motion, but certain things always increased either the power or the number of the motions. Thus, if he stood up, having been seated, the number did not alter, but the force of the blow on the thigh increased remarkably.

"If while standing he elevated and extended the right hand and arm to the shoulder level, instantly the rhythm mounted to two hundred; and when the right arm ceased to act the number fell again speedily to one hundred and sixty.

"When there was no pendulum spasm he could perform with

the left arm any voluntary act not involving the hand, which itself never ceased to twitch; but while the swinging spasms lasted he could execute no volitional act, and the effort to move the limb enormously increased the spasms.

"Excitement and emotion and all forms of electricity added to the force of the motions, but voluntary movements of other limbs increased the number more than the force. Attempts at passive motion, as the effort to fully extend the partially-flexed fingers, cause intense pain in the occiput, just as the effort to overcome rigid gastrocnemii in certain cases gives rise to pain in the dorsal spine. He has the power to stop the spasms by certain manœuvres. If he seizes the left hand with the right and, flexing the left arm, holds it, there is a kind of general spasm; the left hand for a moment seems to struggle with increasing violence; he totters; the face is convulsed; there is horrible pain in the back of the head. Then he gently releases the left arm, which, save for a slight tremor or twitching of the unquiet fingers, remains at rest, and may not move in violent spasm for an hour or more, and is sometimes nearly still for twelve hours.

"He avoids the use of one hand to stop the other, because of the great pain it causes in the head. When he stops the hand with his leg he has little head pain, but it is altogether so unpleasant for him to check it that he rarely does so. When standing, if he wishes to stop the pendulum spasm, he throws the left leg back so as to trip the toe; the arm then falls in as it moves, and he brings the leg forward so as to catch the arm against the thigh, where its own spasm holds it. Then there is a general convulsive movement of the entire body, and the limb is at rest.

"When the arm is hanging quiet at his side, it begins to move if he walks a few steps, or if he lifts the right arm, in which at times—especially after sudden arrest of the spasm—I noticed some large tremor.

"In all of this strange set of symptoms there is no loss of consciousness, no anæsthesia, no ocular trouble or spasm, no aural defect. When he walks long or fast the legs have some disposition to become rigid, but this is an inconstant feature."

Since this graphic account was published, the phenomena presented have undergone some changes. The simple pendulum movement is now replaced by a more complex rotary movement, and which is present only under certain conditions. Thus, when Mr. C. is sitting or standing still, the arm is carried in a semi-flexed position, while the hand and fingers are markedly contracted and in constant vibration. Were it not for this last factor —namely, the vibration—the position would bear some resemblance to that assumed by patients suffering from secondary contracture.

If Mr. C., however, attempts to rise or to walk, the forearm becomes slightly less flexed, whilst the vibration increases both in amplitude and rapidity. If the walk be persisted in, and be it

Fig. 19.

Fig. 20.

ever so slow and gentle, the forearm finally becomes completely extended, and now the entire arm describes a rotary motion, such as is depicted in the upper series of Plate 557, and in Fig. 19.

It will be observed that this movement is one in which the hand, while it is suspended from the shoulder like a pendulum, describes a circle by first sweeping forward, then outward, then backward, and, finally, forward again to its original position. Fig. 19 illustrates the extent of this movement in the outward direction.

If, now, while the arm is performing this peculiar movement, Mr. C. attempts to raise the arm up to the shoulder, the rotary movement increases wonderfully in extent. (*Vide* Plate 557, middle series.) The arm describes a circle, indicated by the arrows in Fig. 20, which, like Fig. 19, is the result of superimposed tracings of the individual pictures. The arm extended, and the hand and fingers contracted, describe a circle in front of the body. The hand sweeps upward and inward, passes the face, sweeps upward and outward, then downward, and then inward again to its original position.

At the same time that this spasm of the left arm is taking place the right arm, too, becomes convulsed. (*Vide* Fig. 20.) It is raised abruptly, and the forearm directed upward, and a series of to-and-fro movements commence, and which examination shows are *synchronous* with the rotary movement of the opposite limb. As the rotating limb rises, the right arm rises; as the former descends, the latter descends. As the left arm sweeps inward and upward, the right arm steadily ascends; as the left arm goes outward and downward, the right arm steadily descends. There is certainly here a curious association of movement.

If, instead of raising the arm to the shoulder, Mr. C. simply sharply flexes the left forearm at the time it is rotating, as in Fig. 21, a series of to-and-fro movements replace the rotary movements. (*Vide* third series, Plate 557, and Fig. 21.) That is, the left arm is thrown violently backward and forward. At the same time the right arm becomes similarly affected, and it, too, is thrown violently backward and forward. As in the previous experiment, the movements are found to be synchronous.

Fig. 21.

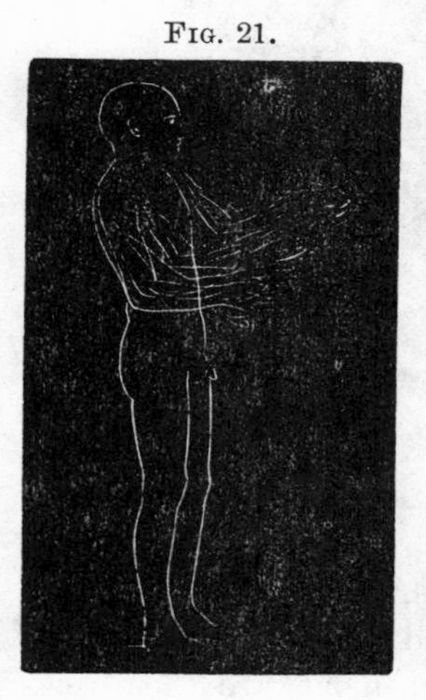

It will be observed that at no time does the right hand become contracted and the fingers "clawed."

Some idea may be gained of the rapidity of these movements when we reflect that the complete cycle of the movement represented in Fig. 19 occupied less than .47 part of a second, while in Fig. 20 the rapidity had so much increased that the cycle occupied only .32 of a second.*

* The intervals of time between the positions of the upper series of Plate 557 were .048 of a second. The entire movement is included in ten pictures.

In Plate 556 further interesting details of this case are illustrated.

If Mr. C. is sitting quietly in a chair, with the arm in the semiflexed position already described, and then lies down with his back and head flat upon the ground, all vibration in the hand ceases. It becomes perfectly quiet, and he can execute with it most exact and delicate voluntary movements. If, however, he now attempts to raise the head, violent tremor at once appears in

FIG. 22.

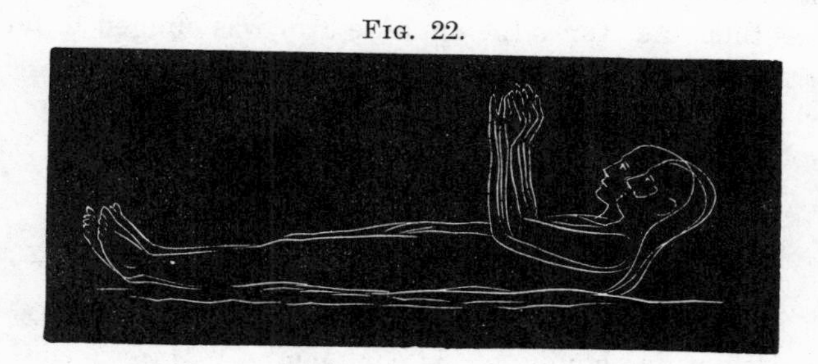

the hand. (See Fig. 22.) This tremor is so violent that the feet and head are affected. It is remarkable that raising of the head is the only movement that provokes the tremor. Raising the right hand or either foot has no effect.

FIG. 23.

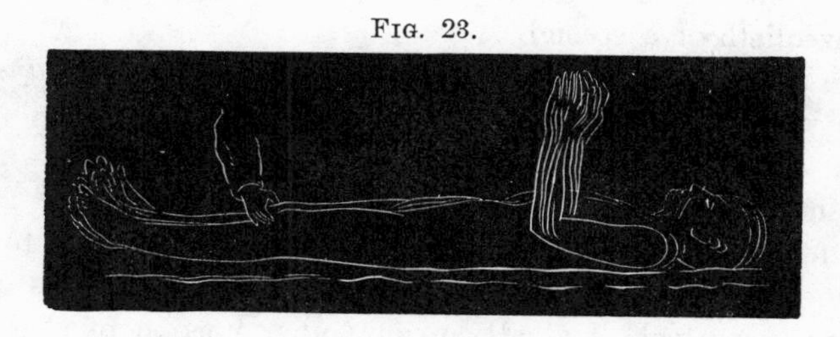

Similarly, if Mr. C. lie upon the ground and the motion be completely arrested, it is again violently excited by striking the patellar tendon (see Fig. 23, and series second of Plate 556) or by attempting to elicit ankle clonus.

The duration of exposure for each individual picture was less than .004 second. The sum total of the nine intervals and ten exposures is (.43 + .04 = .47) equal to .47 of a second. In series B, Plate 557, six positions complete the movement.

It is also a remarkable fact that the tremor so excited, or excited by raising the head, cannot be quieted *unless Mr. C. arises completely from the ground and lies down anew.*

It is not within the province of this monograph to speculate upon this interesting case. However, a point that is of exceedingly great interest is the diffusion of the convulsive movement to other parts of the body, notably to the other arm. This fact is exceedingly suggestive with regard to the spreading of the tremor or clonic spasm in the artificially-induced convulsions. It seems as though at times the opposite arm was obliged to take up an *overflow of nerve-force.*

It should be stated that Mr. C. is still able to arrest the movement of the left arm, at least in part, by seizing it by the right hand, very much in the manner described by Dr. Mitchell.

Wheel-Photographs of Tremors.

Mr. Muybridge having at his command a "Marey" wheel, the writer resolved to test the possibility of photographing tremors. The wheel, which contained eight fenestra, revolved before the camera five times in two seconds, so that in the course of one second twenty exposures could be made, or, in other words, the interval of time between any two exposures was equivalent to one-twentieth of a second.

The first case selected was one of typical *paralysis agitans.* It was a woman, fifty-six years of age, who gave the following history: About four years ago she had intense pain in the ankles and shoulders. She described it as rheumatism. The ankles were much swollen. After the pain subsided, she began to have shaking or trembling, especially in the left arm. It was not at that time marked, if at all, in any other portion of the body. At the time the photographs were made, however, the tremor affected both arms about equally. She complained of burning pain in the shoulders. The position she assumed was typical of an advanced stage of the disease. When standing, the trunk and head were thrown forward, the forearms partially flexed on the arms, and the hands and fingers bent into the "writing hand." The latter was in a constant, regular tremor. When she was asked to move the arms, it was noticed the movements were extremely slow and that the arms were very stiff. When she was asked to walk she showed a marked tendency to fall forward.

The head was not affected by the tremor, and was carried in a rigid, fixed position. The patient never turned her head unless specially asked to do so.

The tremor appeared quite rhythmical, and when the writer first saw the patient it entirely disappeared upon voluntary movement. Later on, and at the time of making the photographs, it ceased to disappear altogether, though it grew much less.

In making the experiment, the patient, as in Marey's experiments on the normal walk, was dressed in a suit of dark-blue tights. On these a white stripe ran along the outside of the arm from the shoulder to the wrist and hand, upon which a bright metallic button was sewed. The legs were similarly striped. The patient was then seated on a chair, and at a given signal she raised both the arm and leg of one side. The result is shown in Figs. 24 and 25.

Fig. 24.

Fig. 25.

In Fig. 24, the patient started to raise the arm a little after the exposures with the wheel had commenced, consequently the stripe down the arm was heavily photographed, and the two bright dots seen at its lower end correspond to the rapidly-vibrating button on the hand. Each successive stripe from below upward is still tipped by a double dot, though the dots appear slightly closer together. The interval between the dots is of course blurred.

Another interesting fact is illustrated by this photograph. The movement consisted in raising the hand to a level with the top of the head. It is noticed that the movement is commenced slowly and with apparent difficulty. The first line above the heavy white stripe is almost in contact with the latter, and cannot be seen except upon close inspection. Thence the intervals between the various lines gradually increase until towards the completion of the movement, when they again appear to diminish. Though this increase of rapidity of movement and subsequent decrease is the same in principle as that which takes place normally, still the slowness of the start and the slowness of the entire movement are remarkable. For instance, the arm sweeps over thirteen intervals before the stripe touches the chin and nose,—that is, counting up to the upper impression of the face, as during the movement the patient threw her head and shoulders backward. The thirteen intervals correspond to $\frac{13}{20}$, or about two-thirds of a second. Therefore it took the patient all of this time to move her hand up to a level of the face, although she was told to move as quickly as possible.

No tremor is noticed in the leg or foot. In Fig. 25 the arm and foot were also raised at the same time and then brought down again. No tremor is noticed in the foot, but both in the ascent and descent of the arm the double rows of dots are plainly seen.

This persistence of the tremor of paralysis agitans during voluntary movement is a matter of considerable importance from the diagnostic point of view. It is generally accepted that one of the chief diagnostic features of the disease is the cessation of the tremor during voluntary movement. This was certainly the case when the patient was first seen, but as the disease progressed the cessation no longer took place, and this is proven by the photograph.

The other case, photographed by means of the "Marey wheel," was a man, aged sixty-five, who had been incapacitated from work for about seven months by reason of a tremor of the right hand. The case appeared to be one of commencing *paralysis agitans*, although some of the symptoms, such as rigidity and festination, were not typical.

The photograph of Fig. 26 was made in exactly the same manner as the preceding ones, with the exception that the patient wore the white stripe upon the arm only. There was a metallic button on the shoulder, at the elbow, and on the hand. He was seated upon a chair and told to rise at a given signal. His right hand was resting on the chair just where the arched piece of the back is joined to the seat. When everything was in readiness he was told to rise. The tremor which had been marked while the hand was resting on the chair gradually became less and less evident, and finally ceased when the erect position was reached.

FIG. 26.

The man is in the act of rising and walking forward. A bright button is fixed upon the hand. The double row of dots indicate the tremor of the hand.

In Fig. 26 the lower row of dots is seen to be double at the beginning of the act, thus showing the presence of tremor. As the man approaches the erect position the dots are gradually less widely separated and finally coalesce. In the case of the elbow, and also of the shoulder, a single continuous row of dots alone is seen.

From the above figures it is very evident that a properly constructed "Marey wheel" offers a valuable and accurate method of studying not only tremors but probably also other forms of abnormal movement.

ADDENDUM.

In Mr. Muybridge's catalogue a number of the clinical plates are incorrectly designated.

Plate 549 should be "Locomotor Ataxia."
Plate 555 should be "Muscular Atrophy of Left Leg."
Plate 558 should be "Stuporous Melancholia."
Plate 559 should be "Partial Paraplegia."

INDEX.

THE LITERATURE OF PHOTOGRAPHY

AN ARNO PRESS COLLECTION

Anderson, A. J. **The Artistic Side of Photography in Theory and Practice.** London, 1910

Anderson, Paul L. **The Fine Art of Photography.** Philadelphia and London, 1919

Beck, Otto Walter. **Art Principles in Portrait Photography.** New York, 1907

Bingham, Robert J. **Photogenic Manipulation.** Part I, 9th edition; Part II, 5th edition. London, 1852

Bisbee, A. **The History and Practice of Daguerreotype.** Dayton, Ohio, 1853

Boord, W. Arthur, editor. **Sun Artists** (Original Series). Nos. I-VIII. London, 1891

Burbank, W. H. **Photographic Printing Methods.** 3rd edition. New York, 1891

Burgess, N. G. **The Photograph Manual.** 8th edition. New York, 1863

Coates, James. **Photographing the Invisible.** Chicago and London, 1911

The Collodion Process and the Ferrotype: Three Accounts, 1854-1872. New York, 1973

Croucher, J. H. and Gustave Le Gray. **Plain Directions for Obtaining Photographic Pictures.** Parts I, II, & III. Philadelphia, 1853

The Daguerreotype Process: Three Treatises, 1840-1849. New York, 1973

Delamotte, Philip H. **The Practice of Photography.** 2nd edition. London, 1855

Draper, John William. **Scientific Memoirs.** London, 1878

Emerson, Peter Henry. **Naturalistic Photography for Students of the Art.** 1st edition. London, 1889

*Emerson, Peter Henry. **Naturalistic Photography for Students of the Art.** 3rd edition. *Including* The Death of Naturalistic Photography, London, 1891. New York, 1899

Fenton, Roger. **Roger Fenton, Photographer of the Crimean War.** With an Essay on his Life and Work by Helmut and Alison Gernsheim. London, 1954

Fouque, Victor. **The Truth Concerning the Invention of Photography:** Nicéphore Niépce—His Life, Letters and Works. Translated by Edward Epstean from the original French edition, Paris, 1867. New York, 1935

Fraprie, Frank R. and Walter E. Woodbury. **Photographic Amusements Including Tricks and Unusual or Novel Effects Obtainable with the Camera.** 10th edition. Boston, 1931

Gillies, John Wallace. **Principles of Pictorial Photography.** New York, 1923

Gower, H. D., L. Stanley Jast, & W. W. Topley. **The Camera As Historian.** London, 1916

Guest, Antony. **Art and the Camera.** London, 1907

Harrison, W. Jerome. **A History of Photography Written As a Practical Guide and an Introduction to Its Latest Developments.** New York, 1887

Hartmann, Sadakichi (Sidney Allan). **Composition in Portraiture.** New York, 1909

Hartmann, Sadakichi (Sidney Allan). **Landscape and Figure Composition.** New York, 1910

Hepworth, T. C. **Evening Work for Amateur Photographers.** London, 1890

*Hicks, Wilson. **Words and Pictures.** New York, 1952

Hill, Levi L. and W. McCartey, Jr. **A Treatise on Daguerreotype.** Parts I, II, III, & IV. Lexington, N.Y., 1850

Humphrey, S. D. **American Hand Book of the Daguerreotype.** 5th edition. New York, 1858

Hunt, Robert. **A Manual of Photography.** 3rd edition. London, 1853

THE LITERATURE OF PHOTOGRAPHY

AN ARNO PRESS COLLECTION

Anderson, A. J. **The Artistic Side of Photography in Theory and Practice.** London, 1910

Anderson, Paul L. **The Fine Art of Photography.** Philadelphia and London, 1919

Beck, Otto Walter. **Art Principles in Portrait Photography.** New York, 1907

Bingham, Robert J. **Photogenic Manipulation.** Part I, 9th edition; Part II, 5th edition. London, 1852

Bisbee, A. **The History and Practice of Daguerreotype.** Dayton, Ohio, 1853

Boord, W. Arthur, editor. **Sun Artists** (Original Series). Nos. I-VIII. London, 1891

Burbank, W. H. **Photographic Printing Methods.** 3rd edition. New York, 1891

Burgess, N. G. **The Photograph Manual.** 8th edition. New York, 1863

Coates, James. **Photographing the Invisible.** Chicago and London, 1911

The Collodion Process and the Ferrotype: Three Accounts, 1854-1872. New York, 1973

Croucher, J. H. and Gustave Le Gray. **Plain Directions for Obtaining Photographic Pictures.** Parts I, II, & III. Philadelphia, 1853

The Daguerreotype Process: Three Treatises, 1840-1849. New York, 1973

Delamotte, Philip H. **The Practice of Photography.** 2nd edition. London, 1855

Draper, John William. **Scientific Memoirs.** London, 1878

Emerson, Peter Henry. **Naturalistic Photography for Students of the Art.** 1st edition. London, 1889

*Emerson, Peter Henry. **Naturalistic Photography for Students of the Art.** 3rd edition. *Including* The Death of Naturalistic Photography, London, 1891. New York, 1899

Fenton, Roger. **Roger Fenton, Photographer of the Crimean War.** With an Essay on his Life and Work by Helmut and Alison Gernsheim. London, 1954

Fouque, Victor. **The Truth Concerning the Invention of Photography:** Nicéphore Niépce—His Life, Letters and Works. Translated by Edward Epstean from the original French edition, Paris, 1867. New York, 1935

Fraprie, Frank R. and Walter E. Woodbury. **Photographic Amusements Including Tricks and Unusual or Novel Effects Obtainable with the Camera.** 10th edition. Boston, 1931

Gillies, John Wallace. **Principles of Pictorial Photography.** New York, 1923

Gower, H. D., L. Stanley Jast, & W. W. Topley. **The Camera As Historian.** London, 1916

Guest, Antony. **Art and the Camera.** London, 1907

Harrison, W. Jerome. **A History of Photography Written As a Practical Guide and an Introduction to Its Latest Developments.** New York, 1887

Hartmann, Sadakichi (Sidney Allan). **Composition in Portraiture.** New York, 1909

Hartmann, Sadakichi (Sidney Allan). **Landscape and Figure Composition.** New York, 1910

Hepworth, T. C. **Evening Work for Amateur Photographers.** London, 1890

*Hicks, Wilson. **Words and Pictures.** New York, 1952

Hill, Levi L. and W. McCartey, Jr. **A Treatise on Daguerreotype.** Parts I, II, III, & IV. Lexington, N.Y., 1850

Humphrey, S. D. **American Hand Book of the Daguerreotype.** 5th edition. New York, 1858

Hunt, Robert. **A Manual of Photography.** 3rd edition. London, 1853

Hunt, Robert. **Researches on Light.** London, 1844

Jones, Bernard E., editor. **Cassell's Cyclopaedia of Photography.** London, 1911

Lerebours, N. P. **A Treatise on Photography.** London, 1843

Litchfield, R. B. **Tom Wedgwood, The First Photographer.** London, 1903

Maclean, Hector. **Photography for Artists.** London, 1896

Martin, Paul. **Victorian Snapshots.** London, 1939

Mortensen, William. **Monsters and Madonnas.** San Francisco, 1936

*__Nonsilver Printing Processes:__ Four Selections, 1886-1927. New York, 1973

Ourdan, J. P. **The Art of Retouching by Burrows & Colton.** Revised by the author. 1st American edition. New York, 1880

Potonniée, Georges. **The History of the Discovery of Photography.** New York, 1936

Price, [William] Lake. **A Manual of Photographic Manipulation.** 2nd edition. London, 1868

Pritchard, H. Baden. **About Photography and Photographers.** New York, 1883

Pritchard, H. Baden. **The Photographic Studios of Europe.** London, 1882

Robinson, H[enry] P[each] and Capt. [W. de W.] Abney. **The Art and Practice of Silver Printing.** The American edition. New York, 1881

Robinson, H[enry] P[each]. **The Elements of a Pictorial Photograph.** Bradford, 1898

Robinson, H[enry] P[each]. **Letters on Landscape Photography.** New York, 1888

Robinson, H[enry] P[each]. **Picture-Making by Photography.** 5th edition. London, 1897

Robinson, H[enry] P[each]. **The Studio, and What to Do in It.** London, 1891

Rodgers, H. J. **Twenty-three Years under a Sky-light,** or Life and Experiences of a Photographer. Hartford, Conn., 1872

Roh, Franz and Jan Tschichold, editors. **Foto-auge, Oeil et Photo, Photo-eye.** 76 Photos of the Period. Stuttgart, Ger., 1929

Ryder, James F. **Voigtländer and I:** In Pursuit of Shadow Catching. Cleveland, 1902

Society for Promoting Christian Knowledge. **The Wonders of Light and Shadow.** London, 1851

Sparling, W. **Theory and Practice of the Photographic Art.** London, 1856

Tissandier, Gaston. **A History and Handbook of Photography.** Edited by J. Thomson. 2nd edition. London, 1878

University of Pennsylvania. **Animal Locomotion. The Muybridge Work at the University of Pennsylvania.** Philadelphia, 1888

Vitray, Laura, John Mills, Jr., and Roscoe Ellard. **Pictorial Journalism.** New York and London, 1939

Vogel, Hermann. **The Chemistry of Light and Photography.** New York, 1875

Wall, A. H. **Artistic Landscape Photography.** London, [1896]

Wall, Alfred H. **A Manual of Artistic Colouring, As Applied to Photographs.** London, 1861

Werge, John. **The Evolution of Photography.** London, 1890

Wilson, Edward L. **The American Carbon Manual.** New York, 1868

Wilson, Edward L. **Wilson's Photographics.** New York, 1881

All of the books in the collection are clothbound. An asterisk indicates that the book is also available paperbound.